The Dynamics of Plant Growth

The Dynamics of Plant Growth

Integrating Morphology, Physiology, and Development

E. David Ford

Emeritus Professor, School of Environmental and Forest Sciences,
University of Washington

OXFORD
UNIVERSITY PRESS

OXFORD
UNIVERSITY PRESS

Great Clarendon Street, Oxford, OX2 6DP,
United Kingdom

Oxford University Press is a department of the University of Oxford.
It furthers the University's objective of excellence in research, scholarship,
and education by publishing worldwide. Oxford is a registered trade mark of
Oxford University Press in the UK and in certain other countries

Published in the United States of America by Oxford University Press
198 Madison Avenue, New York, NY 10016, United States of America

British Library Cataloguing in Publication Data

Data available

Library of Congress Control Number: 2023946600

ISBN 9780192867179
ISBN 9780192867186 (pbk.)

DOI: 10.1093/oso/9780192867179.001.0001

Printed and bound by
CPI Group (UK) Ltd, Croydon, CR0 4YY

Links to third party websites are provided by Oxford in good faith and
for information only. Oxford disclaims any responsibility for the materials
contained in any third party website referenced in this work.

In memory of Rupert Ford.

Preface

Plant growth responds to changes in the environment, whether seasonal, due to weather, or the result of the growth of the plant itself and its effects on both the plant's external and internal environments. Plants cannot escape experiencing changes, but they have multiple control systems, defined by their genetics, that respond to them and so plants are dynamic systems. Defining plants as dynamic systems determines the way growth and survival should be studied and described. The focus in this book is on seed plants: most research has been into gymnosperms and angiosperms.

The way plant growth has been represented has raised difficulties. Descriptive equations of growth became increasingly complex but still failed to capture variation seen in measurements. Mechanistic approaches attempting to define growth, or an important component, as response to variation in the external environment fail if no account is taken of developmental processes.

Over the last few decades research into genetic control of plant development, morphology, and physiology enabled a change in perspective that provided motivation for this book. The growth of a plant is determined by response to variation in the environment: these responses can be seen in developmental processes, direct physiological processes including acclimation, and anatomical and morphological changes including plasticity. These processes are under genetic control and so vary both between and within species.

Chapter 1 describes the process of plant growth as the result of interactions between six groups of processes and the book is organized in four Parts that examine this description.

Part I shows growth to be the result of interconnected processes—distinct in what they control and possibly influenced by different components of the environment and with sequential effects on the growth process. Growth proceeds through accretion and development of modules and plants have processes protecting against, or repairing, damage caused by environmental variation that needs to be taken into account when defining how growth takes place.

Part II focuses on defining control processes. Analogies are made with systems and terms used in control engineering, but plants have unique control systems with their foundations in genetics. Control systems involved in the scheduling of development and physiological processes and apical meristem initiation and development are outlined. Photosynthesis is integrated with protection and recovery from damage as well as the autotrophic process of carbon fixation.

In Part III theories of how morphology and development affect growth are discussed. To explain differences in growth these need to be integrated with processes producing morphological plasticity and physiological acclimation and extended to incorporate foliage display. Genetic variation in morphology and development is continuous, both within species and over time.

Part IV examines effects of changes in the external and internal environment. The effects of interactions between component processes of growth can change over time as structures and their properties change with increase in plant size.

Chapter 11 summarizes the dynamic properties of plants and describes how understanding of them may be used in developing a method for study of growth of a particular plant or circumstance.

This work has benefitted from working and discussions with many scientists. At the Natural

Environmental Research Council laboratory in Edinburgh I worked in a team with Douglas Deans, Ronnie Milne, and Robbie Wilson with particular thanks for the work reported in Chapter 9; and with colleagues Melvin Cannel and Lucy Sheppard with particular thanks for the work reported in Chapter 4. The plants we worked with showed the need to include developmental processes and genetic variation to understand growth.

At the University of Washington I had the great pleasure of working with an interdisciplinary graduate programme of biologists, statisticians, and applied mathematicians. I worked with students from that programme and the then College of Forest Resources, and much of that is referenced here. That interdisciplinary experience confirmed for me the importance of defining biological processes before attempting to make quantitative models and particularly not to favour an approach simply because it was quantitatively tractable or provided an alluring simplification if it did not represent the biological process effectively. This led to the conviction of the importance of describing component processes of growth and how they interact in producing growth, particularly given developments in our understanding of plant processes.

I am grateful to Ross Kiester and to colleagues at the University of Washington for discussions about plants at various times and venues: Toby Bradshaw, Doug Ewing, Tom Hinckley, Takato Imaizumi, Jennifer Nemhauser, Doug Sprugel, and Liz Van Volkenburgh, and particularly her association with maize research reported in Chapter 8. Also for discussions with members of the international Functional–Structural Plant Modelling group, particularly Paul-Henry Cournède, Christophe Godin, Ted DeJong, Véronique Lefort, and Risto Sievänen.

I wish to express my gratitude to Mathew Handley MD and his colleagues at Kaiser Permanente for their diligence and persistence.

I greatly appreciate the work of professionals at Oxford University Press, Ian Sherman, Sylvia Warren, and Katie Lakina, for seeing this book to publication.

I thank my wife Rosemary for her gentle care, unsparing tolerance, generous support, and sacrifice.

Contents

CHAPTER 1

Introduction

1.1 Plant growth is a dynamic process

Human existence depends upon plants and, according to most climate scientists (Intergovernmental Panel on Climate Change 2014), the weather on which plants depend for their growth is changing— so the question of how plants respond to change is of great interest. However, the effects of climate conditions on plant growth cannot be considered as discrete, independent events for each environmental variable. The normal course of plant growth depends upon responses to changes, such as diurnal and seasonal variation and weather events, in which there is partial correlation between environmental variables.

Furthermore, growth requires interaction between different processes, for example photosynthesis, transpiration, and nutrient and water uptake, each of which responds to variation in at least some different environmental conditions. Interactions between processes are influenced by the structure and internal conditions of the plant that change as the plant develops. Intrinsic variation in plant structure and properties combined with the characteristics of environmental variation that interact with them together form a dynamic system determining growth.

In this chapter some problems in analysis of plant growth are outlined. Experimental methods frequently use an analytical framework of *dose→response*, where the effects of different levels of some environmental factor are measured. It has obviously been a highly successful methodology where single, or at least just a few, factors are of interest. For example, analysis of variance type experiments with agricultural species have long been used to define practical rates of fertilizer and continues to be used (e.g. Balkcom et al. 2019) and the technique has been expanded to analyse effects such as landscape on fertilizer effects (e.g. Desta et al. 2022). Ideally experiments require that conditions of other factors than those under investigation should be held constant. Such an approach is difficult to use where influencing factors, and plant responses to them, may change over time.

In this book it is proposed that plant growth should be studied, *explicitly*, as a dynamic system. Dynamic systems have control processes that govern response to variation in conditions. Many plant control processes are genetically determined and, fortunately, research in plant development has established much about how genetics function, which provides a causal basis for understanding responses to change. The plant should be considered as interacting components, continually responsive to, and producing, change in their interactions. The first task in answering questions about plant growth is to define what the important interactions are in a particular situation and how variation in conditions may affect them.

This chapter outlines requirements for defining plant growth as a dynamic system within which the effects of different types of processes and conditions can be integrated, and why the study of dynamics requires a different approach from that frequently adopted in studies that focus on single environmental factors or plant processes. Genetic analyses repeatedly show involvement of multiple genes in controlling growth differences. An important question is whether processes can be defined that may have a dominant influence under some conditions, particularly if these change over time.

The Dynamics of Plant Growth. E. David Ford, Oxford University Press. © E. David Ford (2023). DOI: 10.1093/oso/9780192867179.003.0001

1.2 Difficulties associated with some plant growth theories

Theories of plant growth have frequently been specified in terms of mathematical models and problems have sometimes been recognized due to discrepancies between model predictions and measurements or observations. However, problems also arise when a theory comes to be considered inadequate because it is constructed on a premise found to be incorrect or it does not predict some features that come to be considered important. This is illustrated by attempts to define the trajectory of growth and to explain growth rate as a mechanistic process.

1.2.1 Attempts to define the trajectory of growth

Much initial research into plant growth focussed on defining the trajectory of weight growth in grams of dry weight, g, and then proposing a theory to explain that trajectory. Typically, the time course of relative growth rate, $g\ g^{-1}\ t^{-1}$, over the growing period was calculated and analysed in terms of the amount of leaf produced and calculations of weight increment per unit leaf area, unit leaf rate, $g\ m^{-2}\ t^{-1}$. Development of this research programme is described by Evans (1972).

Increase in weight of annual plants under controlled growing conditions generally follows an S-shaped-type curve, with a first phase of increasing growth rate followed by one of decrease. An important result is that growth in plant weight cannot be described with simple mathematical formulae even when measured in a constant environment. Thornley and Johnson (2000) describe the evolution of functions that led to the logistic equation which describes an S-shaped curve where W is the weight of the plant, μ is an intrinsic growth rate, t is time, and W_f is the final weight of the plant. The proposed process is that in the first phase of growth the size of the plant has a controlling effect—more foliage is produced which in turn can intercept more light resulting in yet more photosynthetic production— and that in this phase the time course of weight follows an exponential increase. $W = W_0 e^{\mu}$: growth rate, R, $dW/dt = \mu W$ increases, and the relative growth rate (RGR; weight $weight^{-1}\ t^{-1}$) is constant.

Representation of the declining phase of growth rate is based on the idea that growth is limited by total available resources. Resources are not represented directly in the equation but by a final weight, W_f, and growth rate declines as that limit is approached so that weight follows the pattern represented by $W = W_f\ (1-e^{-kt})$.

In fitting the logistic growth curve, and developments from it, difficulties have been encountered in both estimating parameters to define growth rates and the point at which a transition between increasing and decreasing phases occurs (Thornley and Johnson 2000). Such problems led to development of equations with more parameters, most notably the frequently used Richards function (Richards 1959), but also to improvement in the data used in estimation. Hunt (1982) made daily measurements of plant weight over the growth span of an annual plant and rather than use equations that implied a particular mathematical form for growth progression he calculated an accurate description of the growth pattern using spline regression from which relative growth rates and points of transition between phases could be estimated.

Following this approach Shipley and Hunt (1996) produced a smoothed pattern of actual weight progression from replicate daily samples for *Holcus lanatus* (Fig. 1.1a) grown from seed in constant, good growing conditions in growth rooms and, from that, calculated the time course of RGR (Fig. 1.1b). RGR of *H. lanatus* was found to increase to an early maximum, decrease slightly, increase to a secondary maximum, and then decline in a non-linear way to a value close to zero. This irregular development of RGR over time, particularly because it occurred in a constant environment, suggests there are internal plant processes that can cause significant changes. Similarly, Hunt and Evans (1980) found differences in patterns of growth between maize genotypes grown in field crop conditions that they could not explain with the available environmental data and concluded that differences should be sought 'in the plants themselves'. On the basis of this type of variation Hunt (1982, Chapter 8) describes plant growth as 'a dynamic, largely continuous, and mathematically form-free process'. The parameters and formulation of equations based on growth as logistic do not provided an accurate description of growth.

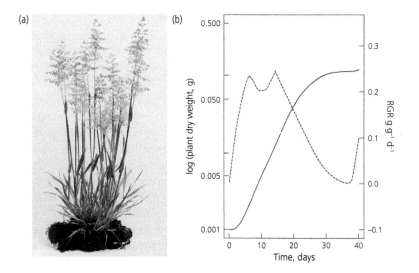

Fig. 1.1 (a) *Holcus lanatus*. (b) Natural logarithm of dry weight (solid line) and RGR (dotted line) plotted against time in days. Part (a) © Jolanta Dąbrowska/Alamy Stock Photo. Part (b) reproduced from Shipley and Hunt (1996) with permission from Oxford University Press.

However, this type of equation is still in use as a description and Paine et al. (2012) review fitting of non-linear growth models and calculating growth rates from them for ecological research.

1.2.2 Environmental influences and analysis based on mechanistic processes

Thornley and Johnson (2000) suggest that if growth equations are considered inadequate it may be more appropriate to seek what they term a mechanistic understanding. A *mechanistic theory* assumes that all the details of a process can be explained by a limited number of physiochemical cause-and-effect relations (Arber 2012, Chapter 11).

A mechanistic approach to the study of plant growth requires selection of particular environmental factor(s). The autotrophic system of plant growth automatically presents solar radiation and photosynthesis and nutrient uptake and utilization as likely candidates for the mechanistic control of growth.

1.2.2.1 Solar radiation and photosynthesis

There has long been an interest in the relationships between solar radiation, photosynthesis, and

growth both in estimating the efficiency of conversion of light energy into growth (e.g. Loomis and Williams 1963) and in how growth may be controlled, or limited by radiation received, for example Black (1964). Analyses have frequently involved calculations of light interception and photosynthetic response and use of models of light interception by the foliage canopy in combination with estimates of photosynthesis response. De Wit (1965) published such a canopy photosynthesis model and models of this type have been developed for practical use, as in estimating crop growth, for example the CERES-maize model (Jones and Kiniry 1986), and typically such models are calibrated against field data for particular use (e.g. DeJong et al. 2011). Problems that have arisen include calculation of the light environment of foliage within the canopy and estimation of photosynthesis response functions. Advances have been made in calculation of the light environment, particularly the nested radiosity approach used by Chelle and Andrieu (1998) that are in use, for example by Wiechers et al. (2011), but have not been generally adopted as a standard method, perhaps in part due to the detailed measurements required of foliage structure.

Variation of rate of photosynthesis (A) has been measured using leaf chamber cuvettes in relation to both light (Q) and CO_2 concentration (C_i), and A/Q and A/C_i relationships have been defined and a theory developed to explain these relationships based on gaseous diffusion. This theory is clearly mechanistic and its mathematical definition uses the constructs of gaseous conductance, notably stomatal conductance, g_s, and mesophyll conductance, g_m, that are considered to affect diffusion of CO_2 from the atmosphere through to the chloroplast stroma (reviewed by Flexas et al. 2008).

A difficulty with a mechanistic theory for photosynthesis is succinctly described by Laisk and Oja (1998):

A lower canopy leaf (maize) was left in the instrument while we went to lunch. When we returned, the low maximum photosynthesis rate of the leaf had increased by a factor of three. The realisation dawned that a leaf is not a constant physical object like a crystal but is able to adjust its responses to environmental conditions.

Photosynthetic capacity of foliage, assessed as the light saturated rate of CO_2 uptake or photosynthetic electron transport from O_2 evolution, varies in relation to such conditions as the developmental stage of plants, age of foliage, and the rate at which products are exported from foliage (Demmig-Adams et al. 2017). Acclimation of photosynthetic rate to environmental conditions (Chapter 3) can involve changes in the structure of the photosynthesis system itself and in foliar vasculature (e.g. Demmig-Adams et al. 2017). These changes are under genetic control and vary between species or ecotypes (Demmig-Adams et al. 2017). Although photosynthesis is obviously essential for plant growth it is not inevitable that its rate determines growth rate (Chapters 3 and 10).

Multiple processes are involved in the control of photosynthetic rate. The supply of water is becoming an increasingly limiting factor to world crop productivity (Morison et al. 2008) and Flexas (2016) indicates that there is an urgent need for *simultaneously* increasing photosynthesis yields and water use efficiency (WUEi) but that very limited success has been obtained towards this objective. Genetic manipulation of stomatal control is possible, but Flexas (2016) states:

the up-to-date knowledge on stomata regulation has also made clear that improving photosynthesis by increasing stomata aperture generally results in decrease in WUEi and increasing WUEi by reducing stomatal aperture results in reduced photosynthesis.

He further points out that at ambient CO_2 and saturating light, photosynthesis is often co-limited by CO_2 and regenerated 5 carbon sugar, ribulose 1,5-bisphosphate (RuBP) which are sources for the enzyme RuBisCo. He suggests that while this is well known to physiologists, the genetic manipulation literature often neglects it.

Flexas proposes that simultaneous improvements of photosynthesis and WUEi can be achieved only through multigene manipulation for all of the limiting factors of photosynthesis. And as conditions vary, both in the physical and chemical environment experienced by the plant and in the condition of the plant itself such as sink strength and translocation, the processes that exert major control of photosynthesis may change over time. Foyer et al. (2017) list a number of contributing processes that may control photosynthesis and may have to be integrated with effects on WUEi. This leads to the problem of finding rational ways to decide on combinations of characters that might be selected for improvement, which is a general problem for phenomics (Tardieu et al. 2017).

A mechanistic theory for photosynthesis might provide a satisfactory explanation for measurements made under controlled conditions and with standardized sampling and measurement procedures. However, such a theory needs to be integrated with a theory for growth and/or plant functioning. A possible connection might be considered through g_m which is of similar magnitude to g_s. g_m varies in relation to many factors (Flexas et al. 2008) such as plant water status, temperature, foliage age, and acclimation status. However, it is not clear from the results of empirical dose-response experiments alone if, or how, a theory for photosynthesis and associated equations, based on gaseous concentrations and conductance, can be extended to incorporate the types of biological variation that may be important.

Korner (2015) suggests that such a carbon-centric approach to the study of plant growth, with photosynthetic CO_2 uptake as the primary driver, does not reflect reality. He suggests this has been the dominant approach and refers to it as 'the classical view' but notes that it is based on restricted conditions used in experiments. An approach of this genre, initially developed with crop growth examples (Monteith 1977, 1994), is the study of radiation use efficiency which proposes that production is proportional to absorbed radiation. This approach requires analysis of the correlation between the accumulated radiation intercepted as a crop grows and total crop dry matter and calculation of a coefficient of radiation use efficiency. It has been severely criticized (Demetriades-Shah et al. 1992) particularly on the grounds that the correlation is described between two running integrated variables; that is, the values do not have the independence required in regression analyses.

Correlation analyses have important difficulties (Box 1.1). With regard to radiation use efficiency Korner (2015) points out that accumulated radiation is related to calendar date and so to plant development, progress of the season, and thus to accumulated temperature and potential evaporation. Consequently the apparent positive correlation does not justify analysis of growth solely in terms of radiation use efficiency.

Box 1.1 Causality and correlation

Scientists frequently work by observing correlations between events or sets of measurements and investigating whether this indicates whether there is a causal relationship between them.

Causation

If C and D are two actual events or occurrences such that D would not have occurred without C, then C is the cause of D. If C, D, E is a finite sequence of particular events such that D depends causally on C, and E on D, then this sequence is a causal chain. One event is a *cause* of another if, and only if, there is a causal chain leading from the first to the second.
(Ford 2000).

A general standard for a causal relationship between components involves transfer of mass or energy, or signal(s) that influence a process, or a change in conditions.

There are problems in interpreting causes from correlations:

1. The problem of effect rather than cause. The time sequence of occurrence needs to be established and how the cause is effected. Causal loops do occur, where C and D have mutual causality, but this needs to be described as such and one event not given priority.
2. The problem of multiple effects—where C causes E then F but E does not cause F.
3. The problem of alternate causes. C_1 occurs and causes E; C_2 also occurs but does not cause E but would have if C_1 had been absent.

Plant growth is the result of multiple interacting processes and this make for difficulty in attributing cause(s) to a growth difference or change.

1.2.2.2 Nutrient uptake and utilization

Thornley (1998) makes a forceful justification for a mechanistic understanding of nutrient uptake and its effects on shoot:root relations based upon representation of source transport and sink processes (Fig. 1.2). The concept of allocation deals with the relative amounts of material that accumulate in different plant parts. Thornley suggests that the two processes of substrate transport and chemical conversion determine allocation and operate according to:

1. substrate sources dependence on shoot and root sizes with possible product inhibition;
2. transport movement down a substrate concentration gradient;
3. linear bisubstrate kinetics at substrate sinks.

Thornley suggests:

There are only two significant types of process in the plant: transport and chemical/biochemical conversion. Both are necessary and sufficient to accomplish allocation. Allocation is the outcome of the processes of substrate supply, transport and utilization.

Thornley illustrates dynamic properties of the model outlined in Fig. 1.2 through simulations of particular circumstances such as defoliation which can result in oscillations of specific growth rate and how this can be damped under different circumstances. An important concern for Thornley (1998) is

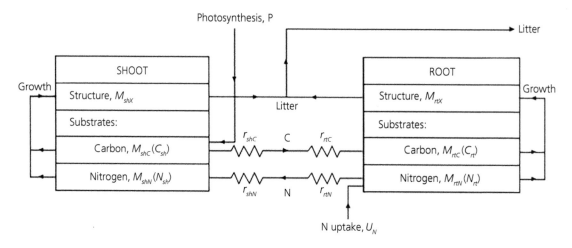

Fig. 1.2 Transport-resistance model of shoot–root allocation using carbon (C) uptake through photosynthesis (P) and nitrogen (N) uptake through the root. Both shoot and root have soluble C and N that can be metabolized to produce growth. Transport of metabolites is controlled by resistances, e.g. $r_{sh}C$ (resistance of carbon transport in the shoot). Transport is calculated as the difference between concentrations in the root and shoot divided by the sum of resistances. Growth rates depend upon the mass of shoot and root components and substrate concentrations. Both P and N uptake are asymptotic with mass assumed due to shading and root competition for nutrients. Reproduced from Thornley (1998) with permission from Oxford University Press.

to exclude teleonomic explanations of the type that plants are efficient or optimal in, say, the utilization of a resource or that there is some 'functional control' by the plant to achieve a particular allocation.

Thornley's approach represents growth and allocation in mechanistic terms as an exclusively physico-chemical process. He notes that input rates of C and N might be replaced by submodels of these processes and of course this could vary by species or conditions. Similarly, resistances may vary. He states:

In this approach substrate sources are connected with transport resistances to substrate sinks where chemical/biochemical conversions take place. The method is termed 'irreducible' because transport and chemical conversion are processes which must take place in order that allocation is accomplished, although how these processes are controlled is debatable.

One difficulty with this particular mechanistic approach is that it is counter to what is actually known about movement of anions and cations within the plant. Both transport through the plant and utilization of anions and cations requires transport through membranes: the plasma membrane between the apoplast and symplast and the tonoplast between the symplast and the vacuole. This

transport is an *active* process, further described in Chapter 4. Active transport concentrates *against* a concentration gradient and this accumulation is essential for the growth process. The requirement is to understand the processes that drive this active process. Taiz et al. (2022) in Chapter 8 of their text on plant physiology note the genetic control of proteins involved in membrane transport. Some 7% of the predicted 25,500 proteins encoded by the *Arabidopsis* genome may be involved in transport.

1.2.2.3 The normative mechanistic theory and its problem

The normative approach to developing mechanistic explanations for plant growth has been to select a physical or chemical variable or condition and to test its effects on growth. This selection might be based on a regression or correlation analysis or on a basic reasoning about plant growth of the type 'Plant growth depends upon photosynthesis, photosynthesis depends upon light, so growth depends upon light' or, as in Thornley's (1998) model, transport and chemical conversions are processes that must take place.

Mechanistic explanations can be deficient where the inference for a supposed causal relationship is actually one of a number that may be considered— as Korner (2015) noted in his critique of radiation use efficiency. Could the analysis of growth proceed through a series of mechanistically based investigations? Multiple variables or conditions may be considered, all of which may have some effect on growth. The question is how a synthesis of these influences may be made. What is required is direct study of interactions between processes and to ensure that interactions are set within the developmental and morphological framework of the plant.

1.3 Genetic control

A fundamental weakness of the mechanistic approach to the study of plant growth is that growth is not exclusively a physical or chemical process. It is a biological process. The plant's genetics determine how growth proceeds and how it responds to, and is affected by, its environment. Furthermore, growth is not a continuously uniform process but is affected by stages in development that are modulated by genetics.

An example of the importance of genetic variation in growth and its component processes is found in the variation in concentrations of metabolites between genotypes that can be used to study the hereditability of metabolism. Li et al. (2016) used recombinant inbred lines (RIL) from parents of two major variants of rice, Lamont (L) the *japonica* subspecies and Taqing (T) the *indica* subspecies, that were known to be divergent in their metabolic profiles. They conducted a quantitative trait locus (QTL) analysis (Box 1.2) of metabolites and some growth characteristics on progeny of a cross between them. A QTL is a region of DNA associated with a specific phenotypic characteristic, in this case based on concentration of metabolites in the plant, or some growth characteristics, that can be identified to specific chromosomes using genetic markers. Li et al. were able to use a population of 280 LT-RIL (restricted fragment length polymorphism [RFLP] type) with 175 markers spanned across all 12 chromosomes. Individuals of the 280 LT-RIL and those of the parent genotypes were grown and then analysed for 512 metabolites and their markers.

Box 1.2 Quantitative trait locus (QTL) and genome-wide association studies (GWAS)

Many characters do not show straightforward Mendelian patterns of dominance and recessive inheritance. Rather they show continuous variation throughout a population, termed polygenic or quantitative inheritance, which can be considered as Mendelian inheritance of many contributing genes. A frequent objective of QTL analysis is to determine whether phenotypic differences in a character are due to variation in a few genes with large effects or many each with small effects.

Two things are required for experimental analysis: organisms that differ genetically for the character of interest, for example contrasting inbred lines of maize; and genetic markers such as single nucleotide polymorphisms of DNA (SNPs) commonly found between genes (Miles and Wayne 2008). The contrasting organisms are crossed, producing an F_1 generation with intermediate phenotypes, and then these F_1 individuals are back crossed with either of the original parents, producing an F_2 generation of recombinant individuals with different combinations of the parental lines. For each individual in the F_2 the phenotype along with genotype of markers that vary between parents is measured.

The F_2 provides a population that is used to examine for the association, or otherwise, of, say, greater phenotype value of a character with particular markers; frequently intervals between markers are considered. Statistical analyses are used to determine if the associations are greater than would be expected from a random distribution of phenotype values relative to the genetic map (e.g. Taylor and Pollard 2010) and where an association is found it is considered to be a QTL.

Research into QTL has evolved into GWAS where multiple parents are selected with particular backgrounds. QTL have been identified that are control genes (e.g. Xiao et al. 2017) and GWAS are an integral component of the functional genomics research programme that deals with the whole structure, function and regulation of genes in contrast to the gene-by-gene approach.

Li et al. found that the majority of metabolic QTL had small to moderate effects and that there were also epistatic effects, i.e. where the effect of one QTL depends upon the presence of others. However, they report the existence of two 'hotspots': regions of a chromosome where QTL are, to some extent, concentrated. These had opposing effects on carbon- and nitrogen-rich metabolites which Li et al. suggest may affect carbon and nitrogen partitioning. They also found that metabolic and plant height QTL were largely distinct but noted that the effective detection of polygenetic effects may require a larger population of offspring than they used.

There are identifiable genetic effects on metabolite composition—Li et al. comment that their result for rice is, in general terms, similar to that found for *Arabidopsis*. However, polygenic inheritance, the occurrence of epistatic effects, and the separation of metabolic and growth QTL suggest that understanding the mode(s) of action requires analysis of the growth processes involved. In the study of plant growth there is unlikely to be a simple dominant gene→growth response effect.

Genetic similarities and differences between plant types and species may be identified (Fig. 1.3). Genetic variability suggests there is not a unique theory of plant growth, universally applicable to all plants, other than in the most general sense, which

is not of great use when asking detailed questions about growth, particularly for a species, cultivar, or ecotype. Many plant species genomes are now available (https://genomevolution.org/coge).

1.4 Overarching theory for plant growth incorporating the dynamics of the process

When considering what may control growth in a particular circumstance there are two linked questions: (i) Are there component processes that have a major effect on growth? This requires defining the developmental processes of a plant's growth as well as understanding sensitivity to different conditions during development. It also involves deciding which components must be studied to provide an explanation for a particular question about growth. (ii) How do these processes interact both with each other and with the environment? This involves defining the causal relations between components

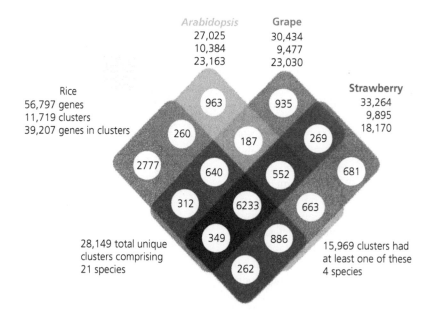

Fig. 1.3 Gene sequencing of the woodland strawberry (*Fragaria vesca*) enabled its comparison with three other species (Shulaev et al. 2011). This Venn diagram represents unique and shared gene families defined as gene cluster among rice (grey), grape (blue), *Arabidopsis* (green), and woodland strawberry (red). 21 species were used to identify 28,149 gene clusters: numbers of genes, clusters, and genes in clusters are shown for each of the four species. Thus for *Arabidopsis* there are 963 gene clusters unique to the species, 260 shared with rice, 187 shared with grape, 640 shared with both rice and grape, 6,233 shared by all four species. Reproduced from Shulaev V., Sargent D. J., Folta K., et al. (2011) with permission from Springer Nature.

and requires making a synthesis of how the system functions.

When considering details of the growth process we might be tempted to conclude that 'everything is connected to everything else' and that the problems of making a synthesis are so large as to be intractable. Fortunately, two features of plant growth assist in the analysis of this problem by enabling a comparative approach to define questions that can be answered.

First, while plant growth is a continuous process it changes with time due to developmental controls and due to the change in a plant's structure and its effect on its own environment. So the problem of making a synthesis that incorporates interactions between processes can be investigated by considering if controls change as growth takes place. This requires that there is some understanding of control systems that define component processes so that their interactions with other processes can be investigated.

Second, there is variability in growth between plant types and between individuals within species. Analysis of this variability, and establishing when uniformity of processes exists, can assist in developing explanations for how growth proceeds in a particular situation. Theories of growth must recognize genetic variability rather than simply assuming uniformity of processes and their interactions for all plants.

Plant growth is a biological process. It is not a chemical process, although chemical synthesis and transformations of compounds are involved. It is not defined by the physical environment, although the environment affects both rates and limits to component processes. The response of plant processes to variation in the environment is determined by the plant's genetics which defines the structure and properties of the plant body and its tissues.

Plant growth is dynamic in two ways. First, the response of a component growth process to a change in conditions may depend upon ancillary conditions or events. Structures and properties of control processes of plants ensure this type of reaction which can change as a plant grows. Second, genetic processes determine how growth takes place; consequently there is inevitably variation, not only between species but also between individuals

within species. The effects of this variation can be seen when growth is studied in different environments and is the result of the dynamics of the genetic system.

The two requirements, that plant growth should be approached as a dynamic system and that multiple processes with different characteristics are involved, are interdependent. When a plant experiences change, whether driven by internal or external factors, it may respond in a number of ways.

An overarching theory for plant growth is required that specifies, in general terms, what might have to be considered in defining the dynamics of plant growth in a particular situation. Plant growth takes place over *time*. It is the result of interactions between the progression of developmental sequences, autotrophic and catabolic physiological processes, and realization of the plant body determined by anatomy and physiology.

Six groups of processes involved in plant growth outline such a theory for examples of growth discussed in this book. Five of these processes are sequential in production of growth—see Fig. 1.4 which shows examples of the structural components of each—and can vary between species and over time as growth takes place. The sixth group of processes are associated with protection against potential damage and repair and occur throughout the plant in various ways.

1. *Nuclear controlled developmental processes initiate, maintain, and regulate growth activity.* Growth is initiated by nuclear activity such as found in seed germination (Taiz et al. 2022, Chapter 17) or the release from dormancy in perennials (Yamane et al. 2021). Throughout the life of a plant developmental processes produce changes in plants' structure, the internal environment, and relationships with the external environment. Chapter 2 provides contrasting examples of the central importance of developmental processes for growth. Chapter 5 describes the role of the circadian clock in maintaining development.
2. *Meristematic regulation of morphology ensures order in plant form.* The control processes that ensure continuation of shoot meristems and regular production of foliage are described in Chapter 5.

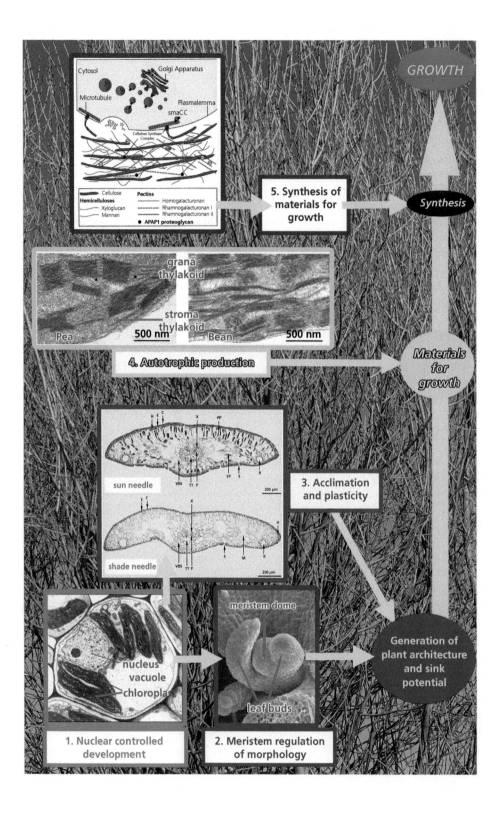

Plant structure has a controlling effect on growth processes: the morphology of a plant not only affects the resources it obtains for making future growth but also, through defining the structure that can be made, determines in part how it can grow. Chapter 2 provides examples of how structure and structural change are important determinants of growth. Control of the continuous development of above ground apical meristems and production of foliar meristems is described in Chapter 5. Examples of the role of structure in controlling growth processes are given for the chloroplast (Chapter 6), shoot development (Chapter 7), including how variability between individuals can occur, foliage (Part III), and the ageing plant (Chapters 2 and 10).

3. *As conditions change there may be acclimation of processes and plasticity of structures.* Acclimation of the photosynthesis process is discussed in Chapters 3 and 6. Morphological plasticity is presented as an integral component of morphological development in Chapters 7 and 8.

4. *Autotrophic processes of photosynthate production and nutrient gain provide essential material for growth.* Control processes regulate the utilization of external energy and chemical sources to provide material in the required form and quantity. Control of photosynthesis is discussed in Chapters 3 and 6 and aspects of nutrient uptake in Chapter 4.

5. *Synthesis of materials in cell growth.* Component processes of synthesis are discussed in relation to control effects of phytohormones in Chapter 5 and cell–water relations and turgor in Chapters 9 and 10.

6. *Protection and repair processes mitigate potential damage to tissues and structures.* Plants maintain a degree of stability in developmental and internal conditions, and study of the dynamics of the processes achieving this indicates that growth must be considered in relation to developmental stability—how its structure and functioning is maintained as the environment and plants themselves change—and stability of internal conditions. The maintenance of developmental stability and its relationship to variability in plant form are discussed in Chapter 7. Conditions within photosynthesizing tissue and the effects that protection and regulatory processes can have on rates of CO_2 uptake are discussed in Chapters 2 and 6 and changes within the ageing plant affecting protection and repair in Chapter 10.

There is nothing unusual about the components given in Fig. 1.4. Its purpose is to indicate that plant growth is the result of interactions between these processes, combined with the sixth group of protection and/or repair. Of course, for some problems the structure represented in Fig. 1.4 may

Fig.1.4 *(Continued)* Five groups of sequential processes involved in plant growth, with examples of associated structural variation that affects the process. Plant architecture and sink potential are generated by (1) nuclear controlled development and (2) meristem regulation of morphology, showing the apical meristem of *Arabidopsis thaliana* with leaf buds on either side. (3) Acclimation of physiological processes and plasticity of morphology are controlled through nuclear processes in response to environmental variation and they affect both the acquisition of resources for growth and development of plant structure: example of sun and shade foliage of *Abies alba*. (4) Materials for growth are the result of autotrophic production. There can be considerable structural variation between chloroplasts of different species that may affect the process of synthesis. (5) An essential primary component of synthesis is the formation of primary cell walls which is the result of a balance between water-induced cell expansion and incorporation of cellulose and hemicelluloses into the expanding wall. Background *Salix* species just prior to foliage bud break. (1) Reproduced from Karen E. Bledsoe, https://people.wou.edu/~bledsoek/about.html. (2) Reproduced from Janoševic D. and Budimir S. *Biologia Plantarum* 50(2):193–197, with permission from the Academy of Sciences of the Czech Republic. (3) Reproduced from Dörken V. M. and Lepetit B. (2018) "Morpho-anatomical and physiological differences between sun and shade leaves in *Abies alba* MILL. (Pinaceae, Coniferales): a combined approach". *Plant Cell and Environment* 41(7):1683–1697, with permission from Wiley and Sons Ltd. (4) Reproduced from Rumak I., Gieczewska K., Kierdaszuk B., et al. (2010) "3-D modelling of chloroplast structure under (Mg^{2+}) magnesium ion treatment. Relationship between thylakoid membrane arrangement and stacking". *Biochemica et Biophysica Acta* 1797:1736–1748, with permission from Elsevier. (5) Reproduced from Jana Verbančič et al. (2018) "Carbon supply and the regulation of cell wall synthesis". *Molecular Plant* 11(1):75–94, with permission from Oxford University Press. Background photo by EDF.

be insufficient—but it is adequate for the examples in this book that illustrate how growth is a dynamic process. The task faced in analysis of growth of a particular situation is to define which processes and interactions have controlling influences. The focus here is on phanerograms because of the emphasis of research on them. There are some 296,462 species of angiosperms and gymnosperms combined (Christenhusz and Byng 2016) and the assumption is made that they have the basic growth components outlined in Fig. 1.4 but that there is variability within those processes and their interactions.

The study of plant growth is necessarily a study of interactions between processes. Integration is achieved in two complementary ways: by defining the developmental sequences involved in growth and how change may affect each in turn as development takes place; and by defining the morphology of the plant that is produced and how that morphology may affect component growth processes and their dependence on environmental factors both external and internal to the plant.

1.5 References

Arber, A. 2012. *The Natural Philosophy of Plant Form.* Cambridge, Cambridge University Press.

Balkcom, K. S., Duzy, L. M., Price, A. J. and Kornecki, T. S. 2019. Oat, rye, and ryegrass response to nitrogen fertilizer. *Crop, Forage and Turfgrass Management*, 5, 180073.

Black, J. N. 1964. An analysis of the potential production of swards of subterranean clover (*Trifolium subterraneum* L.) at Adelaide, South Australia. *Journal of Applied Ecology*, 1, 3–18.

Chelle, M. and Andrieu, B. 1998. The nested radiosity model for the distribution of light within plant canopies. *Ecological Modelling*, 111, 75–91.

Christenhusz, M. J. M. and Byng, J. W. 2016. The number of known plant species in the world and its annual increase. *Phytotaxa*, 261, 201–217.

de Wit, C. T. 1965. *Photosynthesis of Leaf Canopies.* Wageningen, Wageningen University, Pudoc.

Dejong, T. M., Da Silva, D., Vos, J. and Escobar-Gutiérrez, A. J. 2011. Using functional–structural plant models to study, understand and integrate plant development and ecophysiology. *Annals of Botany*, 108, 987–989.

Demetriades-Shah, T. H., Fuchs, M., Kanemasu, E. T. and Flitcroft, I. 1992. A note of caution concerning the relationship between cumulated intercepted solar radiation and crop growth. *Agricultural and Forest Meteorology*, 58, 193–207.

Demmig-Adams, B., Stewart, J. J. and Adams, W. W. 2017. Environmental regulation of intrinsic photosynthetic capacity: an integrated view. *Current Opinion in Plant Biology*, 37, 34–41.

Desta, G., Amede, T., Gashaw, T., et al. 2022. Sorghum yield response to NPKS and NPZn nutrients along sorghum-growing landscapes. *Experimental Agriculture*, 58, e10.

Evans, G. C. 1972. *The Quantitative Analysis of Plant Growth.* Los Angeles, University of California Press.

Flexas, J. 2016. Genetic improvement of leaf photosynthesis and intrinsic water use efficiency in C_3 plants: why so much little success? *Plant Science*, 251, 155–161.

Flexas, J., Ribas-Carbó, M., Diaz-Espejo, A., Galmés, J. and Medrano, H. 2008. Mesophyll conductance to CO_2: current knowledge and future prospects. *Plant, Cell and Environment*, 31, 602–621.

Ford, E. D. 2000. *Scientific Method for Ecological Research.* Cambridge, Cambridge University Press.

Foyer, C., Ruban, A. and Nixon, P. 2017. Photosynthesis solutions to enhance productivity. *Philosophical Transactions of the Royal Society of London. Series B, Biological Sciences*, 372, Article 20160374.

Hunt, R. 1982. *Plant Growth Curves. The Functional Approach to Plant Growth Analysis.* London, Edward Arnold.

Hunt, R. and Evans, G. C. 1980. Classical data on the growth of maize: curve fitting with statistical analysis. *New Phytologist*, 86, 155–180.

Intergovernmental Panel on Climate Change. 2014. *Fifth Assessment Report (AR5).* Cambridge, Cambridge University Press.

Jones, C. A. and Kiniry, J. R. 1986. *CERES-Maize: A Simulation Model of Maize Growth and Development.* College Station, Texas A&M University Press.

Korner, C. 2015. Paradigm shift in plant growth control. *Current Opinion in Plant Biology*, 25, 107–114.

Laisk, A. and Oja, V. 1998. *Dynamic Gas Exchange of Leaf Photosynthesis Measurement and Interpretation.* Canberra, Australia, CSIRO.

Li, B. H., Zhang, Y. Y., Mohammadi, S. A., Huai, D. X., Zhou, Y. M. and Kliebenstein, D. J. 2016. An integrative genetic study of rice metabolism, growth and stochastic variation reveals potential C/N partitioning loci. *Nature Scientific Reports*, 6, 30143.

Loomis, R. S. and Williams, W. A. 1963. Maximum crop productivity: an estimate. *Crop Science*, 3, 67–72.

Miles, C. and Wayne, M. 2008. Quantitative trait locus (QTL) analysis. *Nature Education*, 1, 208.

Monteith, J. L. 1977. Climate and the efficiency of crop production in Britain. *Philosophical Transactions of the Royal Society of London. Series B, Biological Sciences*, 281, 277–294.

Monteith, J. L. 1994. Validity of the correlation between intercepted radiation and biomass. *Agricultural and Forest Meteorology*, 68, 213–220.

Morison, J. I. L., Baker, N. R., Mullineaux, P. M. and Davies, W. J. 2008. Improving water use in crop production. *Philosophical Transactions of the Royal Society London. Series B, Biological Sciences*, 363, 639–658.

Paine, C. E. T., Mathews, T. R., Vogt, D. R., et al. 2012. How to fit nonlinear plant growth models and calculate growth rates: an update for ecologists. *Methods in Ecology and Evolution*, 3, 245–256.

Richards, F. J. 1959. A flexible growth function for empirical use. *Journal of Experimental Botany*, 10, 290–301.

Shipley, B. and Hunt, R. 1996. Regression smoothers for estimating parameters of growth analyses. *Annals of Botany*, 78, 569–576.

Shulaev, V., Sargent, D. J., Folta, K., et al. 2011. The genome of woodland strawberry (Fragaria vesca). *Nature Genetics*, 43(2), 109–116.

Taiz, L., Zeiger, E., Møller, I. M. and Murphy, A. 2022. *Plant Physiology and Development*, Seventh Edition. Oxford, Oxford University Press.

Tardieu, F., Cabrera-Bosquet, L., Pridmore, T. and Bennett, M. 2017. Plant phenomics, from sensors to knowledge. *Current Biology*, 27, R770–R783.

Taylor, S. L. and Pollard, K. S. 2010. Composite interval mapping to identify quantitative trait loci for point-mass mixture phenotypes. *Genetics Research*, 92, 39–53.

Thornley, J. H. M. 1998. Modelling shoot:root relations: the only way forward? *Annals of Botany*, 81, 165–171.

Thornley, J. H. M. and Johnson, I. R. 2000. *Plant and Crop Modelling. A Mathematical Approach to Plant and Crop Physiology*. Reprint of 1990 Oxford University Press edition. www.blackburnpress.com.

Wiechers, D., Kahlen, K. and Stützel, H. 2011. Evaluation of a radiosity based light model for greenhouse cucumber canopies. *Agricultural and Forest Meteorology*, 151, 906–915.

Xiao, Y., Liu, H., Wu, L., Warburton, M. and Yan, J. 2017. Genome-wide association studies in maize: praise and stargaze. *Molecular Plant*, 10, 359–374.

Yamane, H., Singh, A. K. and Cooke, J. E. K. 2021. Plant dormancy research: from environmental control to molecular regulatory networks. *Tree Physiology*, 41, 523–528.

PART I

Variation in plant structure and physiology during growth

Introduction to Part I

Part I provides examples showing that plant growth is the result of interactions between processes that can be affected by changes in developmental processes, plant structure, and the environment. These interactions define a dynamic system.

Chapter 2 focuses on how changes in plant structure and development influence growth and survival. Plants have a modular construction. Growth is the result of accretion of connected modules, and multiple changes in structure and physiology are integral to the development and properties of growth modules. Two examples provide contrasts in terms of plant type and scope of the questions asked but similarity in terms of a definition of structure in terms of modules and their responses to change.

The first is 'How can bristlecone pine live to 5,000 years?' An essential component of the explanation involves the repeated crown die back that occurs and how branches and foliage reiterate from modules following that damage. Other factors are important in providing a complete answer to how it survives, such as control of foliage longevity and possibly sectoralization of xylem—but understanding the structural process is essential.

The second is 'How has production of maize been increased in commercial crops as planting density has increased?' Plant breeding that increased yield has resulted in multiple changes in plant development and morphology, including increased seasonal duration of foliage with a more erect display and change in the timing and synchrony of male and female flowering. The increase in yield over decades can be understood in terms of interactions between these characters but not as the result of one alone.

The natural environment varies continuously due to diurnal and seasonal cycles and as weather systems develop and pass. Chapter 3 illustrates this with the example of radiation and photosynthesis. Components of the effects of environmental variation include: changes in photosynthetic capacity in response to differences in light flux; the induction of photosynthetic capacity following an increase in light flux; and overall reduction in photosynthetic capacity at high radiation. Although photosynthesis is obviously essential for plant growth, the rate of growth is not inevitably controlled by the amount of radiation received. Dynamics of the photosynthesis system must take into account protection against the effects of potentially damaging high light flux and transitions between different light conditions achieved through the structure and properties of enzyme systems.

CHAPTER 2

Effects of changes in development and plant form on growth

2.1 Introduction

Visual variability in plant form is found both between and within species. This has made it challenging to find ways of representing plants so that interaction between their form and function can be analysed. Here plant form refers to the morphology of component structures and how these are organized in the plant body (Bell 2008). An important contribution has been to define growth of the above ground plant as repeated increments of *modules* of growth, with the whole plant viewed as an accumulation of these increments (White 1979). A module comprises a unit of support structure, such as a twig or stem, the foliage it supports, and meristems that can develop to continue growth. Examples show the modular structure of plants and the dependence of growth on their properties.

Properties of modules and the developmental processes that produce them can affect individual plant longevity or reproductive productivity. Plants are genetically programmed to reproduce: but to supply the necessary photosynthate to create reproductive structures they must maintain sufficient foliage. Two examples illustrate that developmental sequences producing a plant's form interact with changes, or differences, in the environment and can produce variable form and growth. Bristlecone pine (*Pinus longaeva*) is a long-lived tree (Salzer et al. 2014) with a process of branch and foliage development that changes the plant's form and maintains growth for up to 5,000 years but with no apparent decline in its capacity for cone production. Maize (*Zea mays*) is a determinate plant with a distinct developmental change in the structures it produces:

a period of vegetative growth terminates in seed formation, grain filling, and then death of vegetative parts. Density of individual plants in maize crops has increased over decades and breeding for increased yield over this period has affected the plant's developmental sequence and structure of modules.

In these examples causal explanations, respectively for great longevity or a marked increase in productivity, need to be based on the interactions between processes of development and the growth, structure, and function of the plant body. Although it may seem attractive to seek explanations based on single plant or environmental characteristics such apparent simplicity does not provide adequate understanding.

2.2 Modular growth above ground

A module is a unit of foliated axis, but variation in plant morphology and development have led to differences in nomenclature, definitions, and views of how modules are organized. Prévost (1967) first used the concept of a jointed section of a plant (*l'article* translated as *module*) defined as:

simple morphogenetic shoot units of determinate growth, constant in their expression, derived one from the other by a sympodial mechanism, the resulting sympodium being linear, branched in one plane, or branched in three dimensions.

Because a foliated axis can be produced by plants in a variety of ways Barthélémy and Caraglio (2007) suggest that Prévost's definition is restrictive and that the term *module* has come to be used to refer to

The Dynamics of Plant Growth. E. David Ford, Oxford University Press. © E. David Ford (2023). DOI: 10.1093/oso/9780192867179.003.0002

the portion of an axis produced by a single terminal axis. Comparison of the structure and growth of the bush *Salvia microphylla* with the tree *Picea sitchensis* illustrates use of the concept of modular growth.

Initial modules of *S. microphylla* grow from a perennial root and stem and have leaves arranged in pairs on opposite sides of the stem (Fig. 2.1) and may terminate with a flower and fruit. This forms a monopodium: a single continuous growth axis (e.g. from base to terminal in Fig. 2.1) that can produce lateral shoots from meristems developing in the axils between a leaf and the stem. A clear example of lateral module growth occurs at the distal end of module B (Fig. 2.1) where modules grow from the axes of each of a pair of leaves, e.g. leaf pair A.

After flowering and fruit development the terminal section of the initial module, shown in (Fig. 2.1), died. New modules continued to be produced from laterals already developed from the initial module. This type of branching is sylleptic: branches grow from the newly initiated lateral axis without the meristem having a rest or delay in growth. A sympodium, referred to in Prévost's definition, is an axis of a plant made up of the bases of successive branches so arranged that it resembles a simple or monopodial axis: the example of cotton (*Gossypium hirsutum*) is given in Chapter 8.

The central, or trunk, axis of *P. sitchensis* is formed through successive increments of annual growth (Fig. 2.2). The apical meristem is active throughout

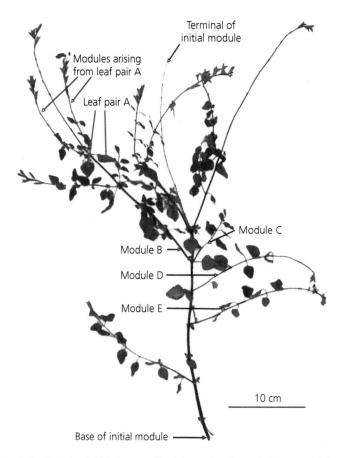

Fig. 2.1 A module of *S. microphylla* with its base initiated at ground level that produced a terminal flower and fruit, now shed. Leaves are opposite and in pairs and new modules may grow from meristems that form in the stem–leaf axil. Production and growth of new modules is not necessarily equal between leaf positions of the same pair of leaves. B is longer than C and single modules, D and E, developed without a complementary module.

Length of height growth module, cm

Number of branches in the whorl

26

31.5

43

41

38.5

13.5

16

8

7.5

5

5

6

7

4

6

5

3

4

Whorl of buds
Terminal module
Whorl of five branches

Inter-whorl branches

Fig. 2.2 A sapling of *P. sitchensis* with nine annual height increments since its seedling stage, given in centimetres at left. Height increments terminate with a whorl of modules (numbers in blue) that grow into branches following the same general pattern of growth as the main axis.

this process (Chapter 7), although overwintering buds form that are a component of the dormancy process. The spatial organization of lateral buds and their extension follows a more rigorous pattern in *P. sitchensis* than *S. microphylla*. Buds form in the axils of needles around the base of the terminal overwintering bud that are spatially regular around the central plant axis in a whorl, which gives rise to regular distribution of new modules that develop into branches. The extension of the main axis module is greater than that of modules of the whorl: a process of apical dominance (Taiz et al. 2022, Chapter 18), though in subsequent years extension of whorl branches can increase. The tree in Fig. 2.2 had been growing on a nurse log (Woods et al. 2021)

on the embankment of the Hoh River, Washington State. Increase in both height increments and number of whorl branches formed is likely the result of increase in size and foliage amount, but with the decline in the last two years possibly the result of competition from neighbours either side. Variation in extension growth and numbers of buds of *P. sitchensis* produced is discussed in Chapter 7.

The clumped arrangement of *S. microphylla* (Fig. 2.3) is the result of there being no single main axis, and that production and growth of modules from the initial module is likely contingent on conditions at the growth site. New module production and growth tends to be greater on the clump periphery. In *P. sitchensis* the regularity of module

Fig. 2.3 Horticultural variety of *S. microphylla*. This plant had been in a position with greater light coming from the left and modules grew more in that direction.

Fig. 2.4 *Picea sitchensis*, Prince William Sound, Alaska. Three examples of excurrent crown structures are marked by arrows, but damage has occurred to other crowns that has disrupted this structure. Photo credit: iStock, Gerald Corsi.

formation and arrangement is the foundation of its typically excurrent crown structure, where the axis is central and other crown parts are regular around it (Fig. 2.4).

An understanding of growth as a complete process requires understanding of variation that can occur in the structure and properties of the module and how this interacts with other characteristics of the species. Examples of properties of modules that regulate effects of environmental variation on growth are described for both *P. longaeva* and *Z. mays*. Interaction between developmental, morphological, and physiological processes in the growth of *P. sitchensis* is discussed in Chapter 7.

2.3 Survival of *Pinus longaeva*

2.3.1 Habitat, environment, and longevity

Pinus longaeva[1] (bristlecone pine) (Fig. 2.5) is restricted to high elevations in California and

Nevada in the United States. It is the plant with the longest individual above ground lifespan: individuals regularly exceed 4,500 years and even up to 4,862 years (Salzer et al. 2009). Following a survey of *P. longaeva* across its distributional range Schulman and Ferguson (1956) conclude that the greatest distribution of old individuals occurred in marginal localities (Fig. 2.5a), and older individuals all had lateral die back of the cambium exposing part of the trunk bare to the xylem (Fig. 2.5c). They suggest that the amount of such die back is a guide to locating oldest individuals.

Pinus longaeva grows slowly. It has been studied extensively to develop tree ring chronologies and elucidate past climate conditions (Salzer et al. 2014; Bruening et al. 2017). Much research into the species and its growth has been conducted in the high desert environment of the White Mountains, California at altitudes between 2,500 and 3,500 m (Billings and Thompson 1957; Ferguson 1968; Connor and Lanner 1989), with maximum mean daily temperatures ~11°C in July (Salzer et al. 2009) and annual precipitation 300–350 mm with most falling

---

[1] *Pinus longaeva* is defined by Bailey (1970). Earlier research conducted at the White Mountains in California refers to the species as *Pinus arista*, which Bailey defines as a species with distribution in the Rocky Mountains of Colorado and northern New Mexico and an isolated group in Arizona, and which also

has the common name of bristlecone pine but generally lives only 1,500 years.

Fig. 2.5 (a) A stand of bristlecone pine growing on dolomite in the White Mountains, California, containing some trees greater than 4,000 years old. Trees vary in amounts of crown die back, with large parts of the complete trunk visible for some. (b) Young bristlecone pine with branches supporting foliage along their complete lengths. (c) A bristlecone pine tree at the same site with no bark remaining on the nearside of the trunk but branches coming from a strip at the right rear. The 1-m stick with 10-cm divisions provides a scale.

as snow (Wright and Mooney 1965). From field measurements and experiments Wright and Mooney conclude the species is predominately found on dolomite soils on which it has greater tolerance for the comparatively low nutrient availability, and so has reduced competitors, and can take advantage of better water relations on this soil than on neighbouring sandstone and granite soils.

2.3.2 Proleptically produced modules and crown reiteration

Although harsh environments, particularly those with lower temperatures, have generally been associated with greater longevity, as Rosbakh and Poschold (2018) show for three herbaceous species, slow growth in a severe environment in itself is not the sole cause of *P. longaeva* great longevity. The form of growth and particularly how this changes over time is important. Typically in conifers the module of growth is an increment of a long shoot and which for *P. longaeva* would be an annual increment (Fig. 2.6a). An actively growing long shoot will extend with each successive module bearing an apical bud that will continue growth to produce a new

module in the following year. Occasionally buds that will subsequently grow into lateral branches are produced, but lateral branching in *P. longaeva* is sparse when compared with other pine species, particularly in young trees, so that foliated long shoots have a bottlebrush appearance (Figs. 2.5b, 2.6a). Successive modules are separated by bud scales or bud scale scars and frequently by slight differences

Fig. 2.6 (a) Long shoots of bristlecone pine showing increments for current and aged one, two, and three years. Needle lengths vary between increments, e.g. those on the module labelled as 3 are markedly shorter than those on the other increments. (b) A short shoot with bud. (c) Short shoots, each supporting five needles which are the visible parts, arranged along an annual increment of a long shoot with the needles of one distal short shoot opened to expose an experimentally induced (see text) terminal bud, marked with an arrow. (d) Two long shoots growing from short shoot buds. Parts (b)–(d) reproduced from Ewers F. (1983) with permission from Canadian Science Publishing.

in needle length (Fig. 2.6a) or colour that enables them to be aged (Ewers and Schmid 1981). Connor and Lanner (1989) report module lengths in the range 2–72 mm but typically with mean values between trees and years in the range 14–29 mm. An interesting feature is that as trees age there is no change in the length of modules produced by long shoots, and both male and female flowers, and cones, continue to be produced.

Foliage of bristlecone pine is produced as needles, generally in groups of five but occasionally four or six, with groups and their supporting shoot (Fig. 2.6b) frequently called a fascicle. These are also called dwarf shoots (Ewers and Schmid 1981), or sometimes short shoots, because their growth in *Pinus* is determinate and restricted to the support of the group of needles (Fig. 2.6c). Connor and Lanner (1989) report long shoot modules supporting between 39 and 50 dwarf shoots. Foliage of *P. longaeva* is extremely long-lived, with a maximum retention time of 45 years compared with other pine species of 2–4 years (Ewers and Schmid 1981). From common garden trials with other *Pinus* species Ewers and Schmid infer both a species genetic effect and an altitude effect, with needle retention being greater at higher altitudes. Generally, individual needles do not abscise with increase in age: rather complete short shoots are lost. Young *P. longaeva* have symmetrical crowns (Fig. 2.5b): the product of symmetrical branch production and growth from the trunk similar to that of *P. sitchensis* (Fig. 2.2). However, a notable feature of mature bristlecone pine is a bush like, fragmented crown frequently below a dead upper section of trunk (Figs. 2.5a, 2.5c). Additional photographs showing a dead upper trunk with lower bush like foliated branches can be seen in Billings and Thompson (1957), Ferguson (1968), and Lanner (2002). The crucial developmental change facilitating great longevity is transition from the extending symmetrical crown of the young tree to the lower bush-like structure that can reiterate continuously. In *Pinus balfouriana*, a related subalpine species, similar crown disruption occurs and that species lives for 2,000–3,000 years.

A structural feature of *P. longaeva* facilitates continued reiteration of foliage in the bush-like crown. A proportion of dwarf shoots can have terminal buds (Fig. 2.6b) that can produce a long shoot (Ewers 1983). These terminal buds are usually not visible due to needles being adpressed to them and so are sometimes called interfoliar buds. Ewers (1983) reports that on trunks 44.3% of dwarf shoots had apical buds compared to 6.6% on lateral long shoots. On lateral long shoots it was the distal dwarf shoots within a module that tended to have interfoliar buds. These buds generally remain dormant, as expected for a dwarf shoot, but if the apex of the long shoot by which they are supported is removed experimentally (Ewers 1983), or by damage, they can become active and produce a new shoot. This is proleptic growth when, in contrast to sylleptic growth, extension occurs after the bud has been dormant for some potential development cycle. The terminal bud of the long shoot in Fig. 2.6d had been experimentally removed as well as all short shoots except those eight-year-olds shown in the figure. In the natural environment the short shoot buds may extend spontaneously, that is without there being removal of the long shoot apex. Ewers suggests extension of dwarf shoots only occurs on older modules, but Connor and Lanner (1987) report younger modules with proliferating interfoliar branches. They indicate that 51% of first-order branches were from interfoliar branches of non-traumatic origin. This type of sprouting is seen both on branches (Fig. 2.7a) and from the main trunk when it will produce new branches (Fig. 2.7b).

Long shoot growth from a dwarf shoot may be initiated by mortality of some branches. Branch die back may be the result of disruption of the root system (Lanner 2002), although a period of drought may have the same effect in causing such die back and is likely to affect branches higher in the tree more readily. Cambial growth in bristlecone pine is positively related on the main trunk or on existing branches to soil moisture in the summer period (Tolwinski-Ward et al. 2011). Branches may develop from dwarf shoots on more proximal positions, i.e. closer to the root system.

The new structure resulting from short-shoot to long-shoot reiteration does not continue development of the old axes but establishes a new and incoherent structure relative to previous growth and, as this process is repeated, it results in a rambling,

(a)

(b)

Fig. 2.7 In bristlecone pine, (a) foliage and branch reiteration arising from a trunk; and (b) repeated reiteration arising along a branch, producing an irregular structure that has been exposed because of the loss of short shoots.

jumbled crown. Both Larson (2001) and Lanner (2002) suggest that within the vascular system there is little lateral movement of water so that individual branches have direct and sealed connections to the root system, and Lanner (2002) suggests this division of the vascular system into distinct sectors enables diminishing proportions of the tree to remain alive. This type of foliage reiteration from short shoots is particularly interesting because it means that as long as there is foliage on the tree there is some potential for branch production. Reiteration in conifers from epicormic buds, vegetative buds that lie dormant for some period beneath the bark of a module, can produce a more regular structure through reiteration but perhaps has less potential for sustaining longevity (Chapter 10).

The survival of *P. longaeva* is a dynamic process (Fig. 2.8), whereby repeated foliage and branch die back is balanced with reiteration of modules and development of new foliage bearing branches that may have cones. It is not certain how the perturbation that causes die back can best be defined, whether as a period of low water potential, damage to a growing apex, or some unidentified process, but a consequence is certainly the die back of mature branches or shoots.

The control process of this dynamic system has anatomical and physiological components: the first stage is the production of apical buds on some short shoots, typically close to the apex of the long shoots. The second stage is the extension of a short shoot apical bud to produce a long shoot that can continue growth from year to year.

2.3.3 Multiple characters contribute to longevity

Repeated cycles of damage and reiteration are not the only characteristic of the species that may contribute to its longevity. The retention of foliage on long shoots for up to 45 years likely gives a sustained return on the carbon investment in

Fig. 2.8 Diagrammatic summary of the processes contributing to the morphological component of longevity of *P. longaeva*.

foliage. Needles of *P. longaeva* appear to be robust for the climate it experiences. Connor and Lanner (1991) report needle cuticle thickness in the range 5–9 μm—almost double the thickness of other timberline species—with no decline in cuticle thickness with increasing age. Although needle length in the range 15–30 mm varied between years, Ewers (1982) found no significant trend with increasing age and reports continuous production of phloem in needles and with advancing needle age. Up to 32 years, the number of living phloem cells remains fairly constant and he suggests the function of needles is maintained. *Pinus longaeva* was found to have stable photosynthetic rates through the growing season (Mooney et al. 1966) even during drought periods: unlike some other treeline species *P. longaeva* does not respond to mid-winter warm periods with an increase in photosynthetic rate, but Schulze et al. (1967) show that respiration rates remain high during winter and they suggest that recovering from that would contribute to the species' slow growth. Lanner and Connor (2001) found no age-related changes in tracheid diameter or accumulation of deleterious mutations and conclude that the concept of senescence, by which they mean age changes in the whole tree, does not apply to this species: there is no natural ageing of meristematic tissue (Chapter 10).

In *P. longaeva* there is not a gradual change in structure: what changes is the origin of long shoots. Due to the structure and function of the module these can be produced from interfoliar buds rather just from buds directly on long shoots. This may occur when the young tree first reaches a height and foliage mass that transpiration load cannot be sustained by water supply, which may result in the die back seen on all long-lived trees. It results in the tree crown developing an irregular shape but one that can be stable for many centuries or decline gradually in total amount of foliage. Tolwinski-Ward et al. (2011) describe episodes of drought occurring at decadal scales in this environment and such events could result in periodic die back. The initial form of the plant is completely disrupted but continues in a stable form for this environment, and this, combined with adaptation to harsh conditions, resistance to mountain pine beetles (Bentz et al. 2017) and low frequency of fire on sites that

have little fuel accumulation (Lanner and Connor 2001), results in great longevity.

The structure and functioning of long shoots and dwarf shoots remain constant over the life of the tree and there appears no decline in reproductive potential of shoots in these irregular-shaped crowns of reduced size. This is obviously a phenomenological description. However, likely explanation of the production of interfoliar buds, particularly their non-uniform distribution, and their growth might reasonably be expected in the framework of hormonal controls of apical control and apical dominance (e.g. Furet et al. 2014).

2.4 Increasing grain yield in *Zea mays*

2.4.1 Plant structure and development

Maize is a monocarpic (semelparous) plant. It produces a sequence of leaves from a single stem and then the terminal tassel of pollen-producing male flowers (Fig. 2.9a). Female flowers initiate as a bud in the axil of a leaf and stem and when maize is grown as a crop typically one or two ears develop from flowers in the central portion of the stem (Fig. 2.9b). After fertilization the grain of the ear is filled by a comparatively stable mass of foliage, with the exception that some leaves towards the base of the plant may senesce and die during the grain filling period. In typical maize cultivation the plant dies when ear growth is complete. Poethig (1994, and references within) describes development of the plant, variation in structure of successive modules (sometimes called phytomers), and some genetic and environmental factors that can result in variation of the plant. Phytomers can be considered as modules comprising a section of stem, i.e. an internode, and its associated leaf sheath and lamina (Fig. 2.9c) connected by a section of tissue, the auricle (Fig. 2.9d). The leaf sheath surrounds the stem of the succeeding phytomer and the lamina curves away from the stem (Fig. 2.9c). The stem of maize consists of an outer circle of vascular bundles below the epidermis and a central core of thin-walled parenchyma cells, the pith, which may contain starch grains. The epidermis and vascular bundles are typically joined by parenchyma with thickened walls (collenchyma and sclerenchyma).

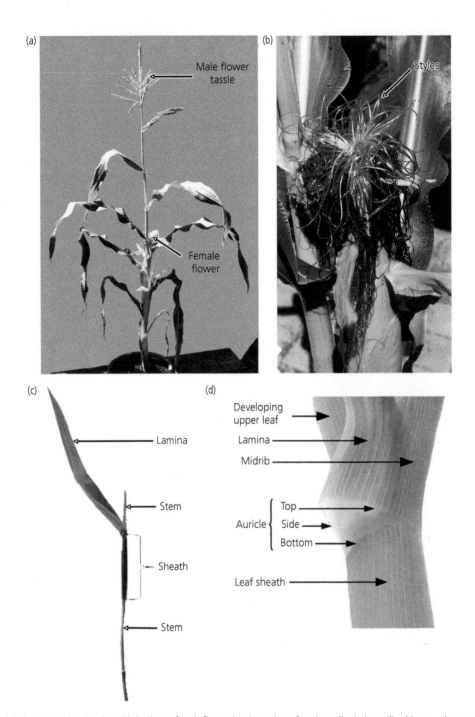

Fig. 2.9 (a) Above ground maize plant; (b) developing female flower showing styles, referred to collectively as silks; (c) vegetative growth module; (d) auricle joining lamina and sheath: the leaf sheath surrounds the stem, and sheaths of successive leaves overlap to form a pseudostem. Part (d) reproduced from Ford E. David et al. (2008) with permission from Elsevier.

2.4.2 A case study of long-term yield improvement

The yield of maize is determined by interactions between (a) the developmental processes of module, stem, tassel, and flower production and (b) the structure of the module and how it affects production and utilization of resources for grain filling. These interactions can be understood by observing changes that have occurred over decades as maize yield has increased. A long-term programme of selection and breeding for greater maize crop yield in the west-central United States Corn Belt (Duvick et al. 2004) was implemented by the commercial company Pioneer HiBred International and genetic gain in maize yield has been continuous since the 1930s.

Breeding started with open-pollinated cultivars but by the 1930s had progressed to the use of inbred parents that were selected by breeders to be crossed and produce hybrids released from the breeding programme for commercial sale.

Whilst selection has been based on performance in yield trials, with no explicit selection for plant structure or development, nevertheless, as yield increased, plant form and some components of plant development changed (Duvick et al. 2004; Lee and Tollenaar 2007). Viable seeds of previously produced hybrids can still be produced, and in direct experiments over a number of growing seasons Duvick et al. (2004) compare yields and some plant characteristics of an historical sequence of 45 hybrids with commercial introduction from 1930 to 2001. They compare yields at each of three planting densities: 10,000 plants ha^{-1}, which they considered approximated to the plant receiving very little competition; 30,000 plants ha^{-1}, typically used in agriculture of the 1930s; and 79,000 plants ha^{-1}, typical of year 2000. Crucially the commercial planting density increased over the period of selection and breeding and yields increased.

Duvick et al. found that oldest hybrids produce their maximum yields at the lower planting density of 30,000 plants ha^{-1}, actually out yielding newer hybrids at that planting density, while new hybrids substantially out yield older ones at 79,000 plants ha^{-1} (Fig. 2.10). The increase in crop yield of newer hybrids is not due to a greatly increased

yield potential per plant when grown at low density. Yield per plant at 10,000 plants ha^{-1} increased from 0.32 kg plant^{-1} for 1930s' hybrids to 0.40 kg plant^{-1} for the year 2000 hybrids. Rather, the greater yield is due to the ability of year 2000 hybrids to produce a minimum amount of grain, about 0.15 kg, on every plant, at higher planting density. The hybrids of older origin when grown at the highest planting density have a higher percentage of barren plants or ears with very few kernels. So an important question is: 'How do newer hybrids produce a greater crop yield at closer spacing when competition between individuals can be expected to be greater?'

The field trials to compare 45 hybrids at three planting densities were replicated over a number of years in the period 1991–2001. An estimate for the overall improvement in yield and changes in measured plant characteristics were calculated by fitting a statistical model to the pooled data so that the differences in Figs. 2.10 and 2.11 are not due to differences in weather: Duvick et al. (2004) provide details. Crop yield close to doubled for the hybrids released around the year 2000 compared to that for hybrids released in the early 1930s (Fig. 2.11a). Unlike improvements in wheat yield over the same time period, yield increases were not related in a major way to change in harvest index, the proportion of plant in grain by weight (Duvick et al. 2004).

Three components associated with development changed. There was: (1) decrease in the duration between anthesis, when pollen is released, and silking when styles grow from the female flower (Fig. 2.11b); (2) decrease in size of the tassel at the apex of the stem (Fig. 2.11c); and (3) extension of leaf greenness at the end of the season, referred to as 'staygreen' (Fig. 2.11d).

Anthesis is the fulcrum of plant development in maize: module production ceases, the terminal male flower produces pollen, and there must be synchronous production of silks (styles that are extensions from the ovary and bearing the stigma) by the female flowers (Fig. 2.9b).

Vlăduţu et al. (1999) describe two factors in the timing of anthesis: timing of the transition of the apical meristem from vegetative module production and the extent of internode growth that takes

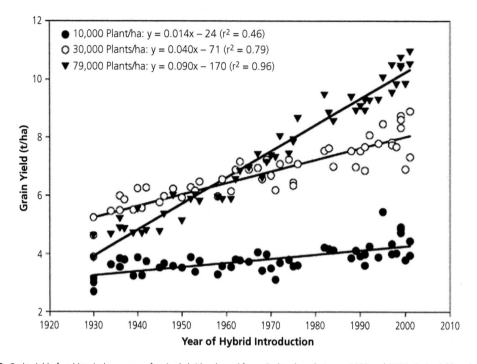

Fig. 2.10 Grain yield of an historical sequence of maize hybrids released for agricultural use between 1930 and 2000. Grain yield per hybrid (x) is regressed on year of hybrid introduction (y) at each of three plant densities (10,000, 30,000, and 79,000 plants ha^{-1}.) Results are based on trials grown in the years 1991–2001, three locations per year, one replication per density. Reproduced from Duvik D., Smith J. S. C., and Cooper M. (2004) with permission from John Wiley and Sons.

place after tassel initiation, i.e. by the internodes that were initiated before the transition of the apical meristem to produce male flowers. Tollenaar (1999) shows the number of leaves produced can be affected by both photoperiod, which affects the phase change of the apical meristem from vegetative module to male flower initiation, and total light received, which can affect internode growth. A direct effect of flowering time on plant structure can be seen in early flowering genotypes which have fewer leaves and are shorter than later flowering genotypes and that number of days-to-anthesis (DA) and days-to-silking (DS) are significantly correlated with leaf number and plant height (Khanal et al. 2011).

Flowering time not only affects the vegetative growth of the plant but also initiates the anthesis-to-silking-interval (ASI) which is important for seed set. Pollen shed generally takes place over a five- to eight-day period, while silks are viable and receptive over a seven- to ten-day period. Generally, a smaller ASI means a greater chance of successful seed set with increased kernel numbers and potentially increased yield, and ASI decreased during the breeding programme described by Duvick et al. (2004). The ear must achieve a biomass threshold for it to produce silks. This is an ear expansion process (Borrás et al. 2009) which can be affected by rate of plant growth and distribution of assimilates. This production of silks depends upon growth of the female flower which can be affected by competition between individuals and for resources within the plant. Uribelarrea et al. (2002) found that increased planting density, and so interplant competition, caused a reduction in silking and lengthening of ASI: there was no lack of pollen despite reduction in tassel size.

There is potential for competition for available resources between tassels and ears at flowering so that reduced tassel size (Fig. 2.11c) is likely to

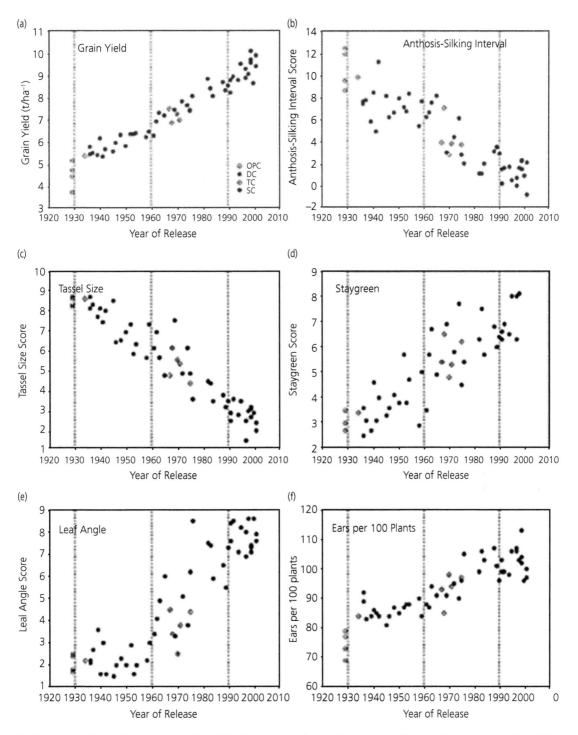

Fig. 2.11 Grain yield (a) and agronomic traits (b–f) of the historical series of maize releases computed from experiments conducted from 1991 to 2001. Dotted vertical lines indicate the distinction between three time periods: 1930–1959, representing the period dominated by double-cross hybrids; 1960–1989, representing the period when the Stiff Stalk and Non Stiff Stalk heterotic groups were developed and single-cross hybrids were introduced; and 1990–2001, representing the modern period dominated by single-cross hybrids. All data are based on three plant densities: 30,000, 54,000, and 79,000 plants ha^{-1}. DC, double-cross hybrids; OPC, open pollinated cultivars; SC, single-cross hybrids; TC, three-way-cross hybrids. Reproduced from Duvik D., Smith J. S. C., and Cooper M. (2004) with permission from John Wiley and Sons

increase potential seed set (Bolaños and Edmeades 1996). The tassels of maize have a high water content, and moisture stress at flowering increases barrenness. Reducing ASI is an important objective of maize breeding in tropical and subtropical regions subject to drought. However, dry conditions can also occur during flowering in the west-central United States Corn Belt, with air temperature reaching 38°C and crop water use in the region of 1 cm a day leading to drying surface and subsoil moisture conditions that may have influenced reduction of tassel size and ASI in hybrids produced in the west-central United States.

'Staygreen' (Fig. 2.11d) is an indication of delayed leaf senescence. Two types are recognized: cosmetic, where the effect is limited to pigment catabolism; and functional, in which the complete senescence process is slowed (Thomas and Ougham 2014). In maize considerable dry matter is fixed during the grain filling period (Lee and Tollenaar 2007), discussed further in Chapter 10, and elongation of that period is likely to increase dry matter gain. Kante et al. (2016) identified quantitative trait loci (QTL) for staygreen in temperate maize and found strong association between genetic markers and staygreen QTL and so identified genomic regions responsible for staygreen in a line used in breeding. They conclude that some alleles may confer staygreen because they increase maximum chlorophyll at flowering, while others because they slow the rate of senescence.

Direct experimental manipulation to increase foliage angle to be more upright increased production in a 1960s maize hybrid (Pioneer 3306) (Pendleton et al. 1968). Visual scores of leaf angle with higher values indicate more erect foliage (Fig. 2.11e), with increase in foliage erectness particularly marked for hybrids produced between the 1960s and 1990s. In a more detailed analysis of foliage Ford et al. (2008) compare the most frequently used hybrids in the central Corn Belt from these two decades, Pioneer 3306 (1960) and Pioneer 3394 (1990) respectively, which had been included by Duvick et al. in their analyses, grown at different planting densities. The length and area of successive leaves change with leaf number from the base of the plant, first increasing to maximum values around the position of the ear (Fig. 2.12) and then

decreasing. Interestingly, leaves of the 1990 hybrid are smaller than those of the 1960 hybrid (Ford et al. 2008).

Leaf curvature changes with leaf position (Fig. 2.12), but it also shows morphological plasticity, with an increase in erectness at closer plant spacing with that of 3394, the 1990s hybrid, consistently more upright. At a planting density of 11,609 plants ha^{-1} foliage of 3394 was more erect than that of 3306. At a planting density of 95,095 plants ha^{-1} foliage of both hybrids was more erect than at 11,609, but the increase was significantly greater for 3394 than 3306 (Ford et al. 2008).

Ford et al. (2008) suggest that an important influence on foliage angle in maize is the interaction between foliage weight and the structure of the auricle, which is the tissue that joins the leaf sheath and leaf blade (Fig. 2.9d). Leaves of the more recent hybrid, 3394, are shorter and weigh less than those of 3306, with a lower estimated torque at the auricle and a significantly smaller auricle angle. Fellner et al. (2003, 2006) propose that the smaller auricle angle is the result of less auricle growth due to reduced sensitivity to the basipetal auxin movement in maize leaves.

Duvick et al. (2004) suggest that more upright leaves improves light utilization at the higher planting densities typically used with later hybrids. An important contribution to this may be a reduction in the competitive shading effect of large on smaller plants so that smaller plants do produce some yield and the numbers of ears in the crop increases (Fig. 2.11f): this process is discussed further in Chapter 8. Edmeades and Daynard (1979) propose that there is a minimum light amount to be received by corn plants below which they do not produce and fill an ear, which relates to the ear expansion process described by Borrás et al. (2009).

The increased uprightness of leaves seen in more recent hybrids marks a reduction in competitive ability of individual plants and is the result of plant breeders making selection for crop yield rather than selection systems based on individual plant yield. Acquisition of resources for growth affected by the environment is obviously of general importance in sequential production of modules, but the amount of resources available during the crucial process of silk production can also be affected by competition

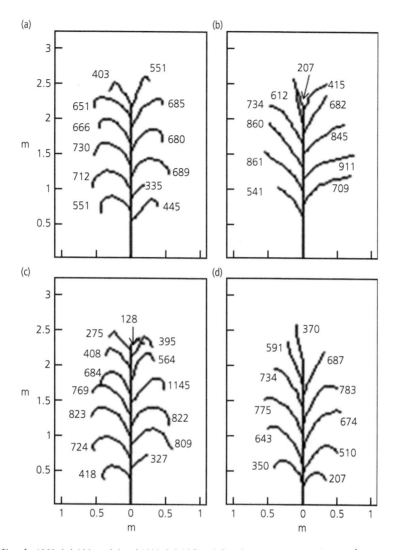

Fig. 2.12 Plant profiles of a 1960s hybrid (a and c) and 1990s hybrid (b and d) each grown at 64,000 plants ha^{-1} (a and b) and 95,095 plants ha^{-1} (c and d). Median size plants of samples of 10 taken shortly after anthesis are shown: numbers beside each leaf indicate their area in square centimetres. In these hybrids some 19 or 20 leaves are produced in total so that the lower ~8 have senesced at the time of these samples. The ear is typically produced at leaf 12. Leaves are represented as alternate, which is the case for the lower half of the plant, after which they tend towards spiral arrangement.

within plants, notably with male flower growth, as well as competition between plants. So there is an important interaction between changes in the development process and in plant form.

2.4.3 Multiple characters affect yield

The reported increase in maize yield achieved through the breeding programme cannot be attributed to a single cause, whether genetic or environmental.

The search for genes that may control characters such as plant height or foliage architecture involves QTL investigations (Box 1.2). From analysis of 5,000 recombinant lines, involving measurement of nearly 1 million plants in eight environments, Buckler et al. (2009) identified numerous small affect QTL for flowering time. Using a multiple family stepwise regression method they identified 36 and 39 QTL that explained 89% of the total variance for DA and DS respectively, whereas 29 QTL explained 64% of the variance in ASI. They comment that ASI appeared more sensitive to genotype-by-environment interactions than other measures of flowering, which agrees with the proposition that silk production is dependent on ear growth. In their QTL meta-analysis Wang et al. (2016) identified meta-QTL regions associated with leaf angle determinants with two genes associated with two of these regions. Aided by the release of the B73 reference genome for maize, genome-wide association studies (GWAS) have identified many yield traits: Xiao et al. (2017) tabulate 94 phenotypic variants.

There are likely additional processes involved not revealed by the changes in traits shown in Fig. 2.11. For example, Sacks and Kucharik (2011) present analyses showing that planting times have become earlier in the United States Corn Belt and the growing season has extended. Li et al. (2015) found increased grain yield in Chinese parental lines in use in the 2000s compared with those of the 1960s and 1980s and attribute this to improved chloroplast structure and greater photosynthetic capacity after anthesis.

With regard to radiation Tollenaar et al. (2017) suggest that over the period 1984–2013 some 27% of an observed yield increase in maize may have been due to accumulated solar brightening during the post-flowering phase of development rather than through the result of changes in agriculture. This is in contrast to the suggestion that while maize growth is source limited prior to anthesis, subsequently it becomes sink limited (Chapter 10).

The majority of important traits in maize are quantitative and controlled by numerous genes with minor effects (Wang et al. 2016). Martinez et al. (2016) present a meta-analysis of QTL studies for grain yield defining 84 meta-QTL distributed across the 10 maize chromosomes and with 74% of QTL individually explaining less than 10% of phenotypic variance. Following analysis of QTL associated with maximum night-time leaf expansion in maize and total leaf length, Dignat et al. (2013) suggest that maximum night-time leaf extension shares much genetic control with growth of other organs where cell wall mechanisms or plant hydraulic properties are likely important. This process and its relationship to carbon gain are discussed further in Chapter 9.

2.4.4 The dynamics of yield improvement

Maize plant growth is a dynamic process. The amount and structure of the plant produced earlier in its life interacts with the timing of the phase change from vegetative to reproductive growth to affect total yield. The genetics of the plant defines four controls of this that have been manipulated through group selection and produced increased yield as planting density has increased (Fig. 2.13). Each control involves response to an environmental condition or signal to change or modify a developmental, morphological, or physiological process.

1. Growth is initiated with the production of modules, and subsequently genetically controlled transition from vegetative to reproductive growth, Control Process 1 affects plant size. An early transition produces plants with fewer leaves.
2. Structure of foliage is affected by Control Process 2 which involves both generation-by-generation change in foliage curvature and plastic response to planting density.

Together Control Processes 1 and 2 determine leaf canopy amount and structure, affecting resources for production of modules and then reproductive growth.

3. Silk growth and control of ASI, Control Process 3, affects seed set. This control involves timing of pollen shed and silk receptivity and ear expansion.
4. Filling of the ear is determined by the function of the canopy following ear set and this can be affected by Control Process 4, leaf senescence.

Ear production is constrained by morphology but within a growing season is affected by resource and water availability at anthesis and silking, as well as during the grain-filling period (Chapter 10). The increase in yield over generations has been due to genetic changes in morphology, developmental processes, and physiology and their interactions in a response to continuous change in planting density.

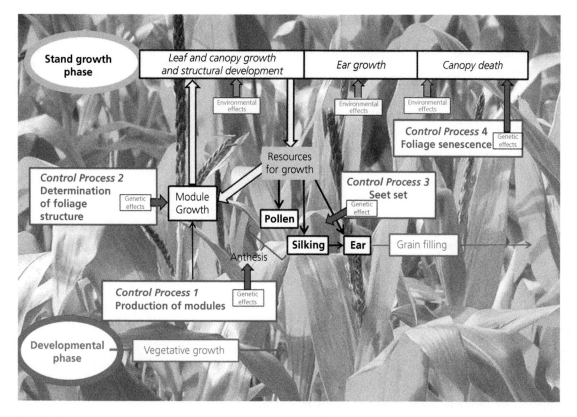

Fig. 2.13 The developmental sequence of maize plants, moving left to right, and influence of environmental effects on yield. Control Processes in red are under genetic control influenced by group selection. Initially Control Processes 1 and 2 are dominant in the formation of the leaf canopy and determine its structure and capacity for dry matter production. Control Process 3 initiates the developmental changes of seed set and then grain filling.

While the component control processes have different genetics, their effects on yield depend upon their interactions in the complete system.

2.5 Discussion

Historically, plant form has not been considered as a determinant of plant growth to the same extent as a plant's physiology, but examples in this chapter show that developmentally produced changes apparent in the structure of the plant are required to understand, respectively, longevity of bristlecone pine and crop productivity of maize. A causal network defines each dynamic response: in the case of *P. longaeva* to damaging events; in the case of *Z. mays* to progressive day length change and its interaction with other environmental conditions.

The structure and properties of modules determine how a causal network for growth functions. The properties of modules, including those of plasticity of supported foliage, regulate some responses to environmental variation; plasticity of foliage angle in *Z. mays* is an obvious example: the angle may be determined during the growth of the foliage, but its effects endure while the canopy exists.

Development and structural changes occur over the life of plants that enable continued plant growth. Within bristlecone pine the process of new module development is sufficiently fecund that long-shoot production can be maintained for millennia. Die back of the crown occurs, but foliage and shoots can be replaced sufficiently to maintain reproduction. Of course what are not understood about this system are details of the causes of the die back and conditions under which it could possibly completely disrupt this system. For example, if the cause is a period of drought resulting in low water potentials that limits growth and causes cell death then what frequency and severity of drought can *P. longaeva* trees sustain? It is a general result that increase in tree height in conifers can result in periods of low water potential in upper foliage sufficient to restrict their growth (e.g. Koch et al. 2004; Meinzer et al. 2008; Mullin et al. 2009; and Chapter 10). With this in mind the transition of individuals of *P. longaeva* from a tree-like form to, effectively, a perpetuating shrub might be expected as a response to drought, although the height differentials are not great. However, this aspect of physiology would need to be investigated to define the control system and its limitations which would be set by the water relations of the plant.

For maize it is particularly interesting that the characteristics of the module to have smaller and more upright leaf blades could be selected from the progeny of inbred lines that were first developed with large spreading, and recurved leaves. This result has a parallel in that of Donohue and Schmitt (1999) and Donohue et al. (2000) who report that inbred lines of the herb *Impatiens capensis* from habitat under woodland cover did not show shade avoidance response in a number of morphological characters that inbred lines from a more open habitat did—where interplant competition within stands of *I. capensis* would be greater. This suggests that the genetic basis for morphological variation that can affect growth in response to a range of conditions exists within the same species.

2.6 References

Bailey, D. K. 1970. Phytogeography and taxonomy of *Pinus* subsection *Balfourianae*. *Annals of the Missouri Botanical Garden*, 57, 210–249.

Barthélémy, D. and Caraglio, Y. 2007. Plant architecture: a dynamic, multilevel and comprehensive approach to plant form, structure and ontogeny. *Annals of Botany*, 99, 375–407.

Bell, A. D. 2008. *Plant Form*. Portland, Oregon, Timber Press.

Bentz, B. J., Hood, S. M., Hansen, E. M., Vandygriff, J. C. and Mock, K. E. 2017. Defense traits in the long-lived Great Basin bristlecone pine and resistance to the native herbivore mountain pine beetle. *New Phytologist*, 213, 611–624.

Billings, W. D. and Thompson, J. H. 1957. Composition of a stand of old bristlecone pines in the White Mountains of California. *Ecology*, 38, 158–160.

Bolaños, J. and Edmeades, G. O. 1996. The importance of the anthesis-silking interval in breeding for drought tolerance in tropical maize. *Field Crops Research*, 48, 65–80.

Borrás, L., Astini, J. P., Westgate, M. E. and Severini, A. D. 2009. Modeling anthesis to silking in maize using a plant biomass framework. *Crop Science*, 49, 937–948.

Bruening, J., Tran, T., Bunn, A., Weiss, S. and Salzer, M. 2017. Fine-scale modeling of bristlecone pine treeline position in the Great Basin, USA. *Environmental Research Letters*, 12, 12014008.

Buckler, E. S., Holland, J. B., Bradbury, P. J., et al. 2009. The genetic architecture of maize flowering time. *Science*, 325, 714–718.

Connor, K. F. and Lanner, R. M. 1987. The architectural significance of interfoliar branches in *Pinus* subsection *Balfourianae*. *Canadian Journal of Forest Research*, 17, 269–272.

Connor, K. F. and Lanner, R. M. 1989. Age-related changes in shoot growth components of Great Basin bristlecone pine. *Canadian Journal of Forest Research*, 19, 933–935.

Connor, K. F. and Lanner, R. M. 1991. Cuticle thickness and chlorophyll content in bristlecone pine needles of various ages. *Bulletin of the Torrey Botanical Club*, 118, 184–187.

Dignat, G., Welcker, C., Sawkins, M., Ribaut, J. M. and Tardieu, F. 2013. The growths of leaves, shoots, roots and reproductive organs partly share their genetic control in maize plants. *Plant, Cell and Environment*, 36, 1105–1119.

Donohue, K., Messiqua, D., Hammond Pyle, E., Heschel, M. S. and Schmitt, J. 2000. Evidence of adaptive divergence in plasticity: density- and site-dependent selection on shade-avoidance responses in *Impatiens capensis*. *Evolution*, 54, 1956–1968.

Donohue, K. and Schmitt, J. 1999. The genetic architecture of plasticity to density in *Impatiens capensis*. *Evolution*, 53, 1377–1386.

Donohue, K., Messiqua, D., Hammond Pyle, E., Heschel, M. S. and Schmitt, J. 2000. Evidence of adaptive divergence in plasticity: density- and site-dependent selection on shade-avoidance responses in *Impatiens capensis*. *Evolution*, 54, 1956–1968.

Duvick, D. N., Smith, J. S. C. and Cooper, M. 2004. Long-term selection in a commercial hybrid maize breeding program. *Plant Breeding Reviews*, 24, 109–151.

Edmeades, G. O. and Daynard, T. B. 1979. The development of plant-to-plant variability in maize at different planting densities. *Canadian Journal of Plant Sciences*, 59, 561–576.

Ewers, F. W. 1982. Secondary growth in needle leaves of *Pinus longaeva* (bristlecone pine) and other conifers: quantitative data. *American Journal of Botany*, 69, 1552–1559.

Ewers, F. W. 1983. The determinate and indeterminate dwarf shoots of *Pinus longaeva* (bristlecone pine). *Canadian Journal of Botany*, 61, 2280–2290.

Ewers, F. W. and Schmid, R. 1981. Longevity of needle fascicles of *Pinus longaeva* (bristlecone pine) and other North American pines. *Oecologia*, 51, 107–115.

Fellner, M., Horton, L. A., Cocke, A. E., Stephens, N. R., Ford, E. D. and Van Volkenburgh, E. 2003. Light interacts with auxin during leaf elongation and leaf angle development in young corn seedlings. *Planta*, 216, 366–376.

Ferguson, C. W. 1968. Bristlecone pine: science and esthetics. *Science*, 159, 839–846.

Ford, E. D., Cocke, A., Horton, L., Fellner, M. and Van Volkenburgh, E. 2008. Estimation, variation and importance of leaf curvature in *Zea mays* hybrids. *Agricultural and Forest Meteorology*, 148, 1598–1610.

Furet, P.-M., Lothier, J., Demotes-Mainard, S., et al. 2014. Light and nitrogen nutrition regulate apical control in *Rosa hybrida* L. *Journal of Plant Physiology*, 171, 7–13.

Kante, M., Revilla, P., De La Fuente, M., Caicedo, M. and Ordás, B. 2016. Stay-green QTLs in temperate elite maize. *Euphytica*, 207, 463–473.

Khanal, R., Earl, H., Lee, E. A. and Lukens, L. 2011. The genetic architecture of flowering time and related traits in two early flowering maize lines. *Crop Science*, 51, 146–156.

Koch, G. W., Sillett, S. C., Jennings, G. M. and Davis, S. D. 2004. The limits to tree height. *Nature*, 428, 851.

Lanner, R. M. 2002. Why do trees live so long? *Ageing Research Reviews*, 1, 653–671.

Lanner, R. M. and Connor, K. F. 2001. Does bristlecone pine senesce? *Experimental Gerontology*, 36, 675–685.

Larson, D. W. 2001. The paradox of great longevity in a short-lived tree species. *Experimental Gerontology*, 36, 651–673.

Lee, E. A. and Tollenaar, M. 2007. Physiological basis of successful breeding strategies for maize grain yield. *Crop Science*, 47, S202–S215.

Li, C.-f., Tao, Z.-q., Liu, P., et al. 2015. Increased grain yield with improved photosynthetic characters in modern maize parental lines. *Journal of Integrative Agriculture*, 14, 1735–1744.

Martinez, A. K., Soriano, J. M., Tuberosa, R., Koumproglou, R., Jahrmann, T. and Salvi, S. 2016. Yield QTLome distribution correlates with gene density in maize. *Plant Science*, 242, 300–309.

Meinzer, F. C., Bond, B. J. and Karanian, J. A. 2008. Biophysical constraints on leaf expansion in a tall conifer. *Tree Physiology*, 28, 197–206.

Mooney, H. A., West, M. and Brayton, R. 1966. Field measurements of metabolic responses of bristlecone pine and big sagebrush in the White Mountains of California. *Botanical Gazette*, 127, 281–297.

Mullin, L. P., Sillett, S. C., Koch, G. W., Tu, K. P. and Antoine, M. E. 2009. Physiological consequences of height-related morphological variation in *Sequoia sempervirens* foliage. *Tree Physiology*, 29, 999–1010.

Pendleton, J. W., Smith, G. E., Winter, S. R. and Johnston, T. J. 1968. Field investigations of relationships of leaf angle in corn (*Zea mays* L) to grain yield and apparent photosynthesis. *Agronomy Journal*, 60, 422–424.

Poethig, R. S. 1994. The maize shoot. In: Freeling, M. and Walbot, A. (eds). *The Maize Handbook*, pp. 11–17. New York, Springer.

Prévost, M. 1967. Architecture de quelques Apoocynacées ligneuses. *Mémoires Société Botanique de France*, 114, 24–36.

Rosbakh, S. and Poschlod, P. 2018. Killing me slowly: harsh environment extends plant maximum life span. *Basic and Applied Ecology*, 28, 17–26.

Sacks, W. J. and Kucharik, C. J. 2011. Crop management and phenology trends in the U.S. Corn Belt: impacts on yields, evapotranspiration and energy balance. *Agricultural and Forest Meteorology*, 151, 882–894.

Salzer, M. W., Hughes, M. K., Bunn, A. G. and Kipfmueller, K. F. 2009. Recent unprecedented tree-ring growth in bristlecone pine at the highest elevations and possible causes. *Proceedings of the National Academy of Sciences of the United States of America*, 106, 20348–20353.

Salzer, M. W., Evan, R. L., Andrew, G. B. and Malcolm, K. H. 2014. Changing climate response in near-treeline bristlecone pine with elevation and aspect. *Environmental Research Letters*, 9, 114007.

Schulman, E. and Ferguson, C. W. J. 1956. Millennia-old pine trees sampled in 1954 and 1955. In: Schulman, E. (ed.) *Dendroclimatic Changes in Semiarid America*, pp. 136–138. Tucson, University of Arizona Press.

Schulze, E. D., Mooney, H. A. and Dunn, E. L. 1967. Wintertime photosynthesis of bristlecone pine (Pinus aristata) in the White Mountains of California. Ecology, 48, 1044–1047.

Taiz, L., Zeiger, E., Møller, I. M. and Murphy, A. 2022. *Plant Physiology and Development*, Seventh Edition. Oxford, Oxford University Press.

Thomas, H. and Ougham, H. 2014. The stay-green trait. *Journal of Experimental Botany*, 65, 3889–3900.

Tollenaar, M. 1999. Duration of the grain-filling period in maize is not affected by photoperiod and incident PPFD during the vegetative phase. *Field Crops Research*, 62, 15–21.

Tollenaar, M., Fridgen, J., Tyagi, P., Stackhouse, P. W. and Kumudini, S. 2017. The contribution of solar brightening to the US maize yield trend. *Nature Climate Change*, 7, 275–278.

Tolwinski-Ward, S. E., Evans, M. N., Hughes, M. K. and Anchukaitis, K. J. 2011. An efficient forward model of the climate controls on interannual variation in tree-ring width. *Climate Dynamics*, 36, 2419–2439.

Uribelarrea, M., Carcova, J., Otegui, M. E. and Westgate, M. E. 2002. Pollen production, pollination dynamics, and kernel set in maize. *Crop Science*, 42, 1910–1918.

Vlăduțu, C., McLaughlin, J. and Phillips, R. L. 1999. Fine mapping and characterization of linked quantitative trait loci involved in the transition of the maize apical meristem from vegetative to generative structures. *Genetics*, 153, 993–1007.

Wang, Y., Xu, J., Deng, D., et al. 2016. A comprehensive meta-analysis of plant morphology, yield, stay-green, and virus disease resistance QTL in maize (*Zea mays* L.). *Planta*, 243, 459–471.

White, J. 1979. The plant as a metapopulation. *Annual Review of Ecology and Systematics*, 10, 109–145.

Woods, C. L., Maleta, K. and Ortmann, K. 2021. Plant–plant interactions change during succession on nurse logs in a northern temperate rainforest. *Ecology and Evolution*, 11, 9631–9641.

Wright, R. D. and Mooney, H. A. 1965. Substrate-oriented distribution of bristlecone pine in the White Mountains of California. *The American Midland Naturalist*, 73, 257–284.

Xiao, Y., Liu, H., Wu, L., Warburton, M. and Yan, J. 2017. Genome-wide association studies in maize: praise and stargaze. *Molecular Plant*, 10, 359–374.

CHAPTER 3

Multiple effects of variation in light on the photosynthesis system

3.1 Introduction

The dominant characteristic of illumination received by plants, in all environments, is that it is variable in intensity. Within foliage canopies large differences occur between intensity from direct sunbeams and when only diffuse radiation is received. This chapter provides examples of the scale and frequency of changes in radiation, and its visible light component, found within a foliage canopy and effects these changes can have on rates of photosynthesis.

Physiological processes determining relationships between light flux and rate of CO_2 fixation can themselves change in response to changes in the environment. Light energy capture involves production of a strongly reducing condition and reactive oxygen within the plant that if not controlled can damage organelles, cells, and tissues. Processes that are an integral part of the complete photosynthesis system mitigate this, and its effects, and so contribute protection against damage but can result in reduced CO_2 assimilation. Also, under conditions of low light flux photosynthetic capacity can decrease and on return to greater light a restoration of photosynthetic capacity is induced; this too is an integral component of the photosynthesis system but can result in delay in the increase of maximum rate of photosynthesis (A_{max}). Considering processes of protection and induction as integrated in the photosynthesis system leads to a requirement for defining the properties that control response to changing conditions. In this chapter measurements of variation of light flux and rates of photosynthesis within a foliage canopy

illustrate the need to take these processes into account.

3.2 Patterns of variability in light received by plants

Radiation flux received directly from the sun depends upon the angle of the sun and its azimuth (direction angle) (Fig. 3.1a). These vary in a regular way throughout a day (Fig. 3.1b), and while there is a consistent daily pattern of increase and then decrease, details of the pattern and values change as the year progresses and sun angle changes.

Calculation for Fig. 3.1 makes assumptions about the transmittance of radiation through the atmosphere. In practice this can vary considerably depending on presence and type of clouds but also on atmospheric particulate matter content and chemical composition (van de Hulst 1979) and generally transmittance is greater for sites at higher elevations (Gates 1980). The site on the Olympic Peninsula (in Washington State, the United States) used for Fig. 3.1 and Figs. 3.2 and 3.3 is 158 m above sea level and 16 km from the Pacific Ocean shoreline. Vegetation was a partly thinned naturally regenerated conifer forest (Ma et al. 2014).

An approximate conversion from total to visible radiation can be made by assuming that 0.45 of radiation is in the photosynthetically active 400–700 nm waveband (photosynthetically active radiation, PAR) and that there are 4.57 μmol m^{-2} s^{-1} Wm^{-2} PAR. Photosynthesis is a quantum process and photosynthetic photon flux density (PPFD) is normally recorded as μmol m^{-2} s^{-1}.

The Dynamics of Plant Growth. E. David Ford, Oxford University Press. © E. David Ford (2023). DOI: 10.1093/oso/9780192867179.003.0003

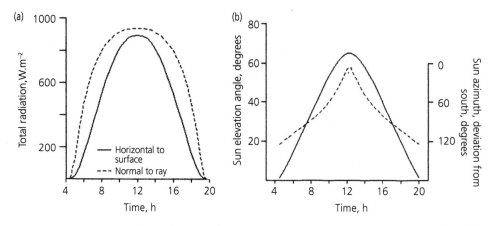

Fig. 3.1 Calculations for 11 June 2009 at a site on the Olympic Peninsula (latitude 47° 49′ 52″ N, longitude 124° 15′ 35″ W), where measurements for subsequent figures were also made. (a) Theoretical incoming radiative flux direct from the sun during daylight hours as received on a horizontal surface (solid line) and a surface normal to the sun's rays (dashed line), calculated from equations in Gates (1980, Chapter 6), assuming a value of direct solar transmittance through the atmosphere of 0.7 and a solar constant of 1.361 kW m^2 (Kopp and Lean 2011; measured by satellite)—a value that includes all radiation, not just visible light. (b) Sun elevation angle (solid line) and azimuth (dashed line); azimuth is represented as degrees deviation from 180°, i.e. south. The sun is at 180° at solar noon, which for this day and site was at 12:16:47 h. Information for sun angle and azimuth can be obtained for a specified latitude and longitude from the website of the United States National Oceanographic and Atmospheric Administration (https://www.esrl.noaa.gov/gmd/grad/solcalc/azel.html).

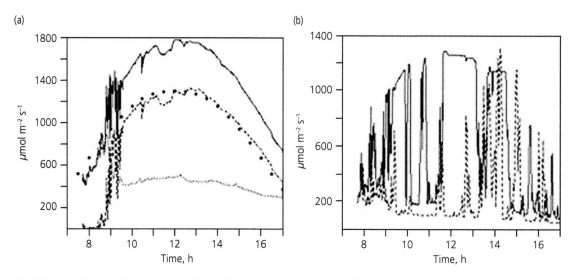

Fig. 3.2 Comparisons of radiative flux received in the visible waveband at a forest on the Olympic Peninsula (latitude 47° 49′ 52″ N, longitude 124° 15′ 35″ W), 11 June 2009, measured with BF3 sunshine sensors in μmol m^{-2} s^{-1} of photons. (a) Above the forest canopy: solid line, total light; dashed line, direct light; dotted line, diffuse light. The sequence of filled circles marks theoretical direct light on a horizontal surface calculated according to Gates (1980) and assuming a solar transmittance of 0.5. (b) Note the difference in ordinate scale from (a). Total light at the positions of two naturally regenerated saplings for the same day as diagram (a): solid line, sapling A with 25–30% canopy openness; dashed line, sapling B with 15–20% canopy openness.

Superimposed on the variation illustrated in Fig. 3.1 is variation due to scattering by the atmosphere and the effects of clouds, which can both intercept light and reflect it onto a site. Two components of radiation received must be considered: that coming directly from the sun and that which has been diffused as the result of scattering by the atmosphere. These can have different temporal patterns as well as generally being received at characteristically different flux. Direct radiation is generally greater than diffuse but can vary over a wider range, and for foliage within canopies direct radiation may only occur for short periods. Furthermore, clouds may obscure the direct beam from the sun but increase the amount of diffuse light so that the amplitude and frequency of change of received radiation depends upon weather conditions and, as far as plants are concerned, their local environment, particularly the occurrence and size of neighbouring plants. Diffuse light is likely received at a lower flux density than direct light but can be more consistent over time.

Different characteristics of variation between direct and diffuse light can be seen in measurements made using BF3 sunshine sensors (Delta-T Devices Ltd) (Ma et al. 2014). These instruments use an array of photodiodes to give values for total visible light and its diffuse light component; direct light is estimated by difference. On a generally clear day total light received on a horizontal surface above the forest canopy increased, as expected, to a maximum at noon and then declined (Fig. 3.2a). Prior to 09:00 h thin cloud resulted in most light being received as diffuse rather than as a direct beam from the sun. When clouds were no longer present from 09:00 h then direct beam light (Fig. 3.2a) followed a very similar trajectory to theoretical values calculated as for Fig. 3.1 but by using a solar transmittance value of 0.5, rather than the 0.7 used by Gates (1980) and for Fig. 3.1a. This suggests that despite the sky appearing blue from mid-morning there were considerable components of the atmosphere, other than clearly visible clouds, that reflected, absorbed, or scattered light. Stanhill and Cohen (2001) note that in the period 1950–1995 there was a continuous reduction in mean measured global irradiance and concluded that the most probable cause is a change in optical properties of the atmosphere. Diffuse light (Fig. 3.2a) from 09:00 h remained relatively constant during the periods of direct light increase and then decrease, though with a small increase and then gradual decline.

Light received by foliage depends upon its surrounding environment; that received over the day by two saplings growing under the forest canopy, but with different amounts of canopy above them, is shown (Fig. 3.2b) for the same day as for light measured above the forest canopy (Fig. 3.2a). The canopy openness for sapling A (solid line in Fig. 3.2b) was estimated as 25–30% using a hemispherical camera (Englund et al. 2000) while that for sapling B was 15–20%.

Light received by both saplings was composed of periods of high and low light, although both their duration and photon flux differ. Over the period 11:30–13:30 h, when sapling A did receive direct light, the total value received was less than that received at the top of the canopy because considerable diffuse light was obscured by the canopy. It is notable that the level of diffuse light, as seen when direct light is not received, is less than the diffuse light received above the canopy. Light received by sapling B is marked by shorter periods of direct light, with generally a lower value than that for sapling A, e.g. compare values at 12:30–13:00 h; this suggests the beam penetrating to sapling B was partially obscured. However, at 14:15 h sapling B did receive the greater value and had some sustained periods of greater light than sapling A, e.g. 14:50–15:10 h.

Intermittent periods of receiving greater relative to lesser light are a characteristic of foliage throughout canopies. Vierling and Wessman (2000) found light to be distributed over time in clusters of high-intensity events, frequently referred to as sunflecks, at each of three heights in a Congolese rainforest, though of course the absolute values varied between them. The variability found in patterns of increase in direct light means that there can be no precise definition of a sunfleck applicable to all foliage canopies and light conditions. Investigators have used different absolute values (Chazdon and Pearcy 1986, 1991) or proportional increases relative to some defined background level (Ma et al. 2014).

On cloudy days when there is no direct beam light there can still be considerable variation in total light received (Fig. 3.3a). An interesting feature of this variation, at least at this site, is its partial regularity, which is likely related to the passage of clouds with different thickness and/or reflectivity. The dominant pattern of variation in Fig. 3.3a analysed using frequency spectra (*spectrum*; R Development Core

Team 2013) is equivalent to a wave-like periodicity of 17.6 minutes. Similar patterns of variation were seen on other cloudy days, although both the frequency and amplitude of variation differed. This pattern of variation seen above the canopy is reflected in patterns at both saplings A and B (Fig. 3.3b); however, the amplitude of variations for A is greater than for B.

The photon flux of diffuse light on the forest floor on an overcast day typically exceeds that of diffuse light on a day with a clear sky, so that on a clear day, unless it is receiving direct beam radiation, light received by a forest floor sapling is *less* than on an overcast day (Fig. 3.4a) (Young and Smith 1983; Ma et al. 2014.) Diffuse light is received from all directions of the sky hemisphere so that

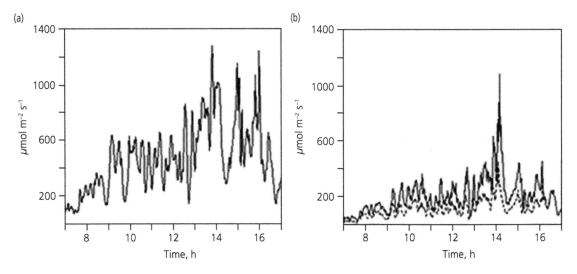

Fig. 3.3 PAR received at the same site as for Fig. 3.2 on 25 June 2009, an overcast day when there was no direct light from the sun. (a) Total light above the forest canopy. (b) Total light at the positions of the same two regenerating saplings represented in Fig. 3.2: solid line, sapling A; dashed line, sapling B.

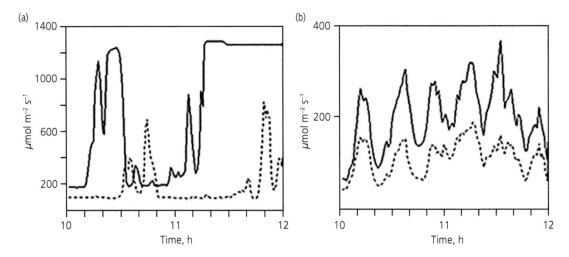

Fig. 3.4 Total PAR received by saplings A (solid lines) and B (dashed lines) for the period 10:00–12:00 h: (a) 11 June 2009, a sunny day; (b) 22 June 2009, an overcast day. Note difference in ordinate scale between (a) and (b).

the pattern of diffuse light received from an over-cast sky (Fig. 3.4b) is similar for the two saplings and reflects the temporal pattern seen above the canopy (Fig. 3.3), though sapling A, which has greater canopy openness, receives more light at all times. The range of variation in diffuse light on these overcast days tends to fall within the range that may be expected to produce an increase or decrease in rate of photosynthesis.

3.3 Seasonal and diurnal variation in potential rate of photosynthesis

Using leaf chambers and gas exchange instruments (e.g. Thornley 2002; Kyei-Boahen et al. 2003; Bellucco et al. 2017) the rate of photosynthesis, A, can be characterized by a well-established curvilinear relationship with light flux, Q, the A/Q curve (Lambers et al. 1998). The measurement technique involves exposing foliage in the chamber to saturating light, e.g. 1,500 μmol m^{-2} s^{-1}, for a sufficient time to obtain stable values of A and then decreasing Q by steps and recording when a new stable value of A is obtained at each step. Conditions within leaf chambers such as temperature, vapour pressure deficit, and CO_2 concentration can be controlled and estimates of stomatal conductance, g_s, and leaf internal CO_2 concentration, C_i, can be obtained for different conditions and used in the interpretation of differences between A/Q curves. The rate of photosynthesis is influenced by environmental factors other than light, which has led to development of models for foliage photosynthesis (Farquhar et al. 2001) that include parameters for gaseous diffusion as well as light processes.

The non-rectangular hyperbola equation (Lambers et al. 1998) can be fitted:

$$A = \frac{\varphi \cdot Q + A_{max} - \sqrt{\dfrac{(\varphi \cdot Q + A_{max})^2}{-4 \cdot \varphi \cdot Q \cdot \theta \cdot A_{max}}}}{2\theta} - R_d \tag{1}$$

where A_{max} is the maximum rate of photosynthesis, θ is convexity of the curve, φ is the apparent quantum efficiency, the initial slope of the linear part of the curve, and R_d is respiration in the dark.

Using this technique Ma et al. (2014) surveyed variation in rates of photosynthesis and respiration of naturally regenerated saplings of two shade-tolerant species, *Tsuga heterophylla* and *Abies amabilis*, throughout a growing season (mid-May to late September) in a partially thinned forest that gave a range of light conditions for saplings on the forest floor.

Saplings were selected in a stratified random sampling method according to differences in canopy cover above them. Shading proportions for each sapling were calculated from measurements of light received over the day of a sample relative to that received above the canopy using both total and diffuse light, identified respectively as Shade$_{TOTAL}$ and Shade$_{DIFFUSE}$. Measurements were generally made at intervals of four to five days and were repeated five times during the day between 10:00 and 19:00 h.

A multivariate regression tree for the complete data with minimization based on sums of squares is shown in Fig. 3.5. This calculation forms a series of nested clusters of the samples by repeatedly dividing the data to minimize dissimilarity within clusters. The major division accounting for most variation is between measurements up to and including 19 August (left-hand branch of Fig. 3.5) being *lower* than those after that date. Within this earlier part of the season, when photosynthetic rate was lower, measurements then divide by leaf temperature above and below 26.6°C (node 1): the 39 measurements with leaf temperature >26.6°C (mean 29.7°C) had a higher respiration rate and lower A_{max} than all other measurements. Interestingly, higher temperature rather than either amount of shading or g_s is the major divider of samples within the first part of the season. For leaf temperature <26.6°C (node 2), a Shade$_{TOTAL}$ of 0.055 condition of the sapling divides the data into 46 saplings in very shaded conditions and 107 in less shaded conditions which then divide on g_s of 0.286 mol H_2O m^{-2} s^{-1} (node 3): 98 saplings with g_s <0.286 and 9 with g_s >0.286 and which have the greater A_{max}. Thus, for saplings in ≥0.055 Shade$_{TOTAL}$ light, stomatal closure may restrict CO_2 assimilation during the first part of the season even when leaf temperature is not >26.6°C.

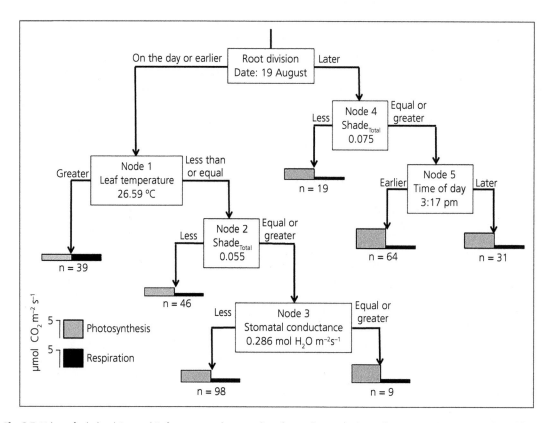

Fig. 3.5 Values of calculated A_{max} and R_d from a season-long sampling of naturally regenerating saplings growing in an 88-year thinned forest partitioned into a nested tree of similarity groups. Each node marks a significant bifurcation in the data. Saplings were shade-tolerant *Abies amabilis* and *Tsuga heterophylla*. There were no significant differences between these species. Shade$_{Total}$ is the proportion of total light received at the sapling over the day of sampling relative to measurement above the forest canopy. This survey was made with saplings receiving <30% above canopy light with 'shade' foliage (Stenberg et al. 1998). Reproduced from Ma Z., Behling S., and Ford E. D. (2014) with permission from Oxford University Press.

After 19 August (right-hand branch of Fig. 3.5), when A_{max} is generally greater, the first division is Shade$_{TOTAL}$ of 0.075 (node 4) separating 19 measures at lesser light and A_{max}. A_{max} was greater on more open sites and among those 95 measurements a division occurs based on diurnal pattern of A_{max} (node 5), in which A_{max} is lower in the late afternoon than earlier in the day. Shade$_{TOTAL}$ divisions for reduced A_{max} were <0.05 prior to 19 August and <0.075 after 19 August which may reflect differences in absolute amount of total light received between those periods.

Figure 3.5 provides analysis of photosynthesis capacity, as defined by A_{max}; it is not based on instantaneous measures of ambient photosynthesis.

It reveals a tension between conditions limiting photosynthesis under greater radiation flux and those limiting due to lesser radiation flux. Generally, measurements of A_{max} prior to 20 August were lower when radiation could be expected to be, and generally was, greater. In that period higher temperature, >26.59°C, is associated with reduced photosynthesis potential and greatest values of respiration. When temperature was <26.59°C foliage that was more shaded (Shade$_{TOTAL}$ <0.055) had lower A_{max}. In this first period reduced g_s is also associated with less photosynthetic capacity in less shaded foliage—there has been extensive research into the role of stomatal closure in the control of transpiration and photosynthesis during water stress (e.g. Buckley 2005).

In the latter part of the season, when A_{max} was generally higher, saplings under greater shade had lower A_{max}, but the decline in photosynthesis capacity in the late afternoon in less shaded saplings is not associated with changes in temperature or g_s but is the time of day when light flux is decreasing.

3.4 Responses to changes in light

The response of foliage to differences in light has to be considered for changes involving response to high light, which potentially may cause damage to photosynthesis tissue, and for change that may increase photosynthesis in a generally low light condition. Both are affected by the conditions under which foliage exists.

3.4.1 Response to potentially damaging or photosynthesis rate reducing high light flux

Reduction in the rate of photosynthesis under conditions of high radiation is extensively reported in the literature, although this has not always been taken into account by scientists investigating growth in relation to radiation received.

The photochemical process of photosynthesis (Taiz et al. 2022, Chapter 9) can potentially be damaging to photosynthetic tissue and there are physiological processes that protect against such damage or repair it. The capture of light energy mediated by chlorophyll in the grana thylakoids generates an electron stream, initiated at the chlorophyll–protein complex termed Photosystem II (PSII) and which produces nicotinamide adenine dinucleotide phosphate (NADPH) utilized in the CO_2 fixation process of the Calvin–Benson–Bassam (CBB), or dark cycle. The photochemical process of PSII releases protons into the thylakoid lumen that drives adenosine triphosphate (ATP) production and also produces oxygen. However, if the production and movement of electrons from PSII becomes unbalanced with more being produced than utilized then high-energy radicals become available for other reactions, particularly the formation of reactive oxygen species (ROS) (molecules with the same molecular structure but charged differently) that can damage organic molecules. This is discussed in greater detail in Chapter 6.

Photochemical processes can proceed more rapidly than the dark cycle which requires gaseous diffusion and a network of enzyme chemistry. Maintaining an electron flow from PSII requires recycling of electron acceptors from dark cycle reactions. Consequently, there is the possibility of ROS being produced and causing damage. Under high light flux the photochemical process can potentially exceed that required for rate of renewal of compounds utilized by the dark cycle.

Three physiological processes can contribute to mitigation of this potential damage: photoprotection by pigments associated with chlorophyll; damage to the PSII reaction centre which reduces electron flow but is subsequently repaired; and photorespiration—the result of the oxygenase property of RuBisCo—which maintains activity of components of the dark cycle process and so, to some extent at least, may contribute to maintenance of electron flow. These three processes need not be exclusive of each other and reduction in g_s may also affect concentrations of CO_2 and O_2 in foliage.

3.4.1.1 Photoprotection

When chlorophyll is excited by light its energy may be used in the photochemical reactions of photosynthesis, called photochemical quenching, or dissipated by transfer to other pigments, and transformed to heat, called non-photochemical quenching. Photochemical quenching requires that PSII reaction centres are open. These centres are chlorophyll–protein structures and photodamage occurs when the D1 membrane protein of PSII is degraded due to oxidative damage (Zavafer et al. 2015), discussed in Chapter 6. Estimation of the distribution between quenching processes involves measuring the condition of PSII reaction centres.

Chlorophyll fluorescence is first measured in dark-adapted foliage and then in foliage exposed to a short burst, e.g. 500-ms pulse, of white light at very high flux (10,000 μmol m^{-2} s^{-1}; Genty et al. 1989), causing transient closure of the centres and producing a maximum fluorescence yield. From such measurements the proportion of open PSII centres in a leaf can be calculated (Genty et al. 1989) along with the concurrent non-photochemical quenching—respectively indicative of light utilized in photosynthesis light reactions and dissipating, typically as heat. If it is not quenched in one way or another then energy of excited chlorophyll may result in damage.

Using this technique Watling et al. (1997) demonstrate potential photoprotection through non-photochemical quenching of three tropical rainforest understory species, *Alocasia macrorrhiza*, *Castanospra alphandii*, and *Alpina hylandii*, in response to sunflecks. Three pigments are involved in non-photochemical quenching (Taiz et al. 2022, section 9.9): violaxanthin (V), antheraxanthin (A), and zeaxanthin (Z). In high light V is converted enzymatically into Z via the intermediary A. Binding of Z to light-harvesting chlorophyll antenna molecules is thought to cause conformational changes that lead to non-photochemical quenching (Chapter 6).

A step change in light from 8 to 800 μmol m^{-2} s^{-1} for 15 minutes increased non-photochemical quenching in the three species (Watling et al. 1997), which rose to around a proportion of 0.8 within 2–3 minutes, after which it increased more slowly. Photochemical quenching declined close to zero but then increased gradually over ~12 minutes to around 0.2 and they attributed this gradual increase to induction of the photosynthesis system, discussed later in this chapter, section 3.4.2, and so greater electron transport through the photosystem.

Field measurements indicated that for all three species PSII photochemistry was very sensitive to changes in PPFD (Fig. 3.6). The effective quantum yield of photosynthesis declined with the onset of the sunfleck and remained below the pre-sunfleck level for some time after it had passed (Fig. 3.6). Pigment conversion of V to A+Z occurred in *A. macrorrhiza* and *C. alphandii* during a sunfleck, but it was more rapid and more of the total pool was converted in *A. macrorrhiza* (71%) than *C. alphandii* (43%) and following the end of the sunfleck *A. macrorrhiza* retained its high level of A+Z for at least 40 minutes whilst in *C. alphandii* the pool declined within 25 minutes to close to pre-sunfleck levels. Full recovery of photochemical quenching occurred by the end of the day.

3.4.1.2 Damage and repair to PSII

The term photoinhibition is applied to a reduction in photosynthetic rate that is recovered when light returns to lower levels; this can be applied to operation of a PSII damage and repair cycle. Photodamage refers to more sustained damage that does not repair, for example blanching and loss of foliage. Photoinhibition is not restricted to conditions of high temperature but can occur when the electron transport system from PSII is slowed (reduced photochemical quenching) or interrupted, e.g. under low CO_2 fixation rates under cold conditions (e.g. Slot et al. 2005).

PSII breakdown and repair is a continuous process and occurs at all light levels but increases as light increases. Repair involves movement within the chloroplast of a damaged reaction centre from the grana thylakoid, where it is associated with chlorophyll light energy collection, to the stroma thylakoid membrane. There the D1 protein is degraded and replaced and the reaction centre returned to the grana thylakoid (described in Chapter 6).

3.4.1.3 Photorespiration

In this process O_2 rather than CO_2 is combined with ribulose-1,5-bisphosphate (RuBP): the oxygenase reaction of RuBisCo. An increase in temperature results in reduced solubility of CO_2 relative to that of O_2, so increasing the potential for the oxygenase reaction. The oxygenase reaction process is frequently referred to as RuBisCo having made a 'mistake'. However, this reaction maintains the dark cycle and so utilizes products of electron flow. The oxygenase reaction is the first in a series of reactions involving transfers of metabolites to peroxisomes and mitochondria in what is termed the C_2 cycle (Chapter 6). Glycolate, a product of the oxygenase reaction, is transferred to peroxisomes, where the enzyme glycolate oxidase contributes to the conservation of fixed carbon and return of glycerate to the chloroplast and dark cycle. Graham (1993) suggests this process was important for the evolution of land plants during a period when atmospheric O_2 was increasing and notes that charophytes, the likely algal ancestors of land plants, also have this enzyme whereas other algae do not. So it may be

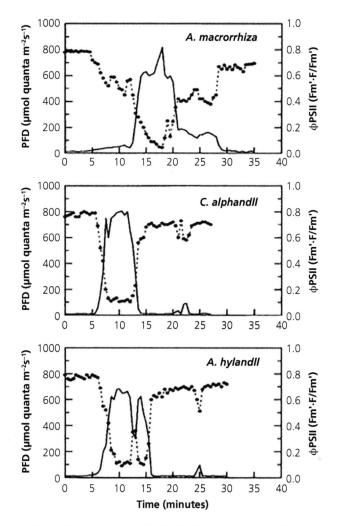

Fig. 3.6 Changes in PPFD (solid lines) and effective quantum yield (dotted lines) before, during, and after sunflecks in *A. macrorrhiza*, *C. alphandii*, and *A. hylandii*. Each plot represents a single plant growing in the understorey. Reproduced from Watling J. R., et al. (1997) with permission from CSRIO.

inappropriate to refer to the oxygenase reaction of RuBisCo as a 'mistake'.

High light flux under field conditions may be accompanied by high leaf temperature. For the diptocarp species *Shorea leprosula*, Leakey et al. (2005) found leaf temperatures in response to sunflecks under field conditions to be some 6°C higher than air temperatures and reaching 38°C. The optimal leaf temperature for CO_2 fixation was 29.1°C. Non-photochemical quenching and proportion of open PSII reaction centres increased markedly with rising

temperatures above 30°C. They exposed plants to a sequence of simulated sunflecks each of 10 minutes at 539 μmol m⁻² s⁻¹ separated by 1 minute at 30 μmol m⁻² s⁻¹ and at both 38 and 28°C. The maximum rate of photosynthesis in high light periods was 43% lower at 38°C. However, although CO_2 fixation was severely inhibited at higher temperature the absorbed energy was effectively dissipated and no photodamage occurred. Leakey et al. contrast this with the condition found by Kitao et al. (2000) where seedlings were exposed to a PPFD of 1,600

$\mu mol\ m^{-2}\ s^{-1}$ and photodamage, i.e. a decline in CO_2 fixation that did not recover within 7 h, which Kitao et al. (2000) found for plants grown in either 10 or 5% of full sunlight before exposure to 1,600 μmol $m^{-2}\ s^{-1}$ but not when raised under open conditions.

3.4.2 Photosynthetic response to high light from a background of low light

When light is increased to foliage receiving low light CO_2 fixation increases gradually: induction is the general term used to describe this increase. An important component is the reactivation of the RuBisCo enzyme, which can become structurally inhibited, and this requires ATP and the operation of the enzyme RuBisCo-activase (rca) that exists in close association with RuBisCo.

An investigation of induction was made with saplings of *A. amabilis* growing in different levels of ambient light on the forest floor. Foliage was subjected to a step change of 1,500 $\mu mol\ m^{-2}\ s^{-1}$, equivalent to a strong sunfleck, in a leaf chamber (Ma et al. 2014). For saplings receiving mean ambient light flux of 145 $\mu mol\ m^{-2}\ s^{-1}$ prior to the step change induction progressed (Fig. 3.7a). Increase in photosynthesis rate is not instantaneous but follows a smooth curve to an asymptotic maximum. This is a frequently reported pattern of induction and can be described by the equation:

$$A = A_{max} * \left(1 - e^{(-t/\tau)}\right) \qquad (2)$$

where A is photosynthesis rate, t is time since the step change (typically in seconds), A_{max} is the final asymptotic value of A, and τ is the time constant and represents the time it takes to reach $1-1/e \approx$ 63.2% of the final asymptotic value. Time constants can vary both between species and between plants growing in different environments. Saplings of *A. amabilis* in this ambient light had mean $\tau = 112.3$ s, with generally only a small increase in g_s, <10%, and no indication of increase in concentration of CO_2 in the substomatal cavity (C_i,) in relation to this slight increase in stomatal opening.

The rate of increase in photosynthesis declines exponentially as light increases from a low value which is compatible with there being a reduction in the number of RuBisCo sites that remain available to be activated with increase in time. This in turn is compatible with a slower and less complete reactivation of RuBisCo sites for a step change in light of 600 rather than 1,500 $\mu mol\ m^{-2}\ s^{-1}$ that produced greater τ and reduced A_{max}—and that regeneration of RuBP proceeds in concert with activation of RuBisCo. The condition of there being a maximum value for A reflects sufficient ATP supply both to provide active sites with the phosphorylated precursor (RuBP) so that carboxylation occurs and to constantly reactivate sites that may become closed. Activation along with the comparable deactivation of reaction sites due to blocking by sugar phosphates occurring as light decreases is a pair of reactions that provides a dynamic response in photosynthesis.

Different patterns of induction were found for saplings growing in lesser ambient light flux. For foliage receiving mean ambient light flux of 74.00 $\mu mol\ m^{-2}\ s^{-1}$ increase in A (Fig. 3.7b) is best described by a two-component process:

$$A = p_1{}^* \left(1 - e^{(-t/\tau1)}\right) + p_2{}^* \left(1 - e^{(-t/\tau2)}\right) \quad (3)$$

where p_1 and p_2 are the proportions of A_{max} accounted for by the two processes, 1 and 2 respectively, and τ_1 and τ_2 are the respective time constants. The first term on the right side of equation (2) represents a faster process. For shoots with this type of induction, mean $\tau_1 = 11.25$ s, mean $\tau_2 = 184.0$ s, and $p_1 = p_2$.

Interpretation of this two-part induction can be based on the laboratory studies of Seemann et al. (1988) with *Alocasia macrorrhiza* and Woodrow and Mott (1989) with *Spinacia oleracea*. Both studies made similar measurements of photosynthesis following the step increase combined with measurements of enzyme activity and cell chemistry using freeze clamp techniques. They observed: (i) an initial rapid change in concentrations of dark cycle metabolites, particularly a decrease in 3-phosphoglyceric acid (PGA) and an increase in RuBP pool sizes, suggesting a rapid ATP- and NADPH-driven regeneration but that is limited in magnitude because, while the first process is rapid, it stops quickly, i.e. its maximum is reached at ~80 s (Fig. 3.7b); (ii) a second phase of light-activated release of RuBisCo from inhibition which is slower than that of the process

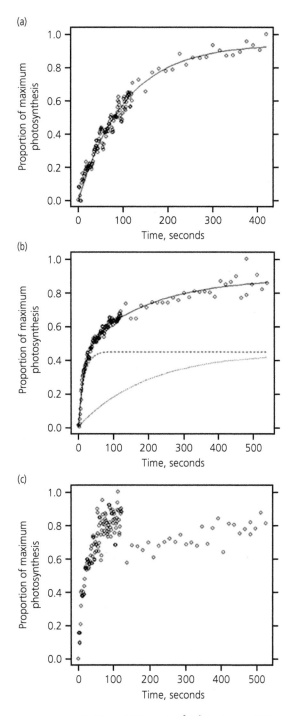

Fig. 3.7 Response of photosynthesis to a step change in light to 1,500 μmol m^{-2} s^{-1}. For foliage of *Abies amabilis* receiving ambient light of (a) 145, (b) 88, and (c) 21 μmol m^{-2} s^{-1}. Sampling was at 1-s intervals for 120 s, then at 15-s intervals. Photosynthesis is represented as proportional value of the maximum measured. For (a), the fitted equation, the solid line is represented by equation (2); for (b) the individual components of equation (3) are represented by the dashed line for the fast component and by the dotted line for the slow component and their sum by the solid line. Reproduced from Ma Z., et al. (2014) with permission from Oxford University Press.

found at higher light, i.e. τ_2 = ~184 s rather than τ_2 = 112.3 s.

This implies that some potentially active RuBisCo sites remain available under the intermediate light flux condition, i.e. they are not in an inhibited condition but can react rapidly with RuBP once ATP and NADPH increase. From their biochemical and enzymatic analyses Seeman et al. (1988) calculated that some 55% of possible RuBisCo sites were available when the step increase in light was made. The slower component process may indicate a lower availability of chemical moieties at this level of ambient light.

For shoots at the least ambient light flux, mean = 43.61 μmol m^{-2} s^{-1}, an overshoot of A occurred (Fig. 3.7c), with a maximum value 80–100 s after the initiation of the step increase in light, followed by a decrease. This can be interpreted as the saturation of unused RuBisCo sites but insufficient supply of substrate. These induction trials, when considered together, indicate the importance of considering activation of RuBisCo together with supply of precursors as components of dynamic response.

Carmo-Silva and Salvucci (2013) suggest that when RuBisCo deactivates under low light RuBP levels generally exceed the concentration necessary to saturate RuBisCo and that light activation of RuBisCo may provide a process that maintains metabolite levels under changing light and thus avoid lags in the resumption of CO_2 assimilation that might be caused by a slow rate of substrate regeneration. However, whilst that may apply to saplings with ambient light of ~139 μmol m^{-2} s^{-1} it seems less likely to apply at lower ambient light levels.

Ernstsen et al. (1997) distinguish three components in the induction response:

1. RuBisCo activation by RuBisCo activase.
2. Regeneration of precursors to the dark reaction. CO_2 fixation requires a molecule of RuBP. Following the production of two molecules of 3-phosphoglycerate from CO_2 and RuBP, a series of enzymatic reactions are required to generate RuBP for further carboxylation (Taiz et al. 2022, Chapter 10). To achieve a sustained increase in rate of photosynthesis the rate of functioning of the dark cycle as a whole must be increased and this requires ATP and NADPH from the light reactions.
3. Stomatal opening which may take up to one hour, but see Vico et al. (2011) for a more detailed analysis and species differences.

3.5 Simulation of effects of variation in light on photosynthesis

Induction trials with *A. amabilis* (Fig. 3.7) were obviously made with a major increase of light flux. A further question relevant for sapling growth in shaded conditions is how foliage may respond to more gradual and smaller changes. The sequence of light values received by sapling B on an overcast day (Fig. 3.4b, lower line) was used to illuminate foliage of *A. amabilis* and photosynthesis was measured. A general pattern of parallel increases and decreases in light and photosynthesis is seen (Fig. 3.8a).

Important differences in this sequence become apparent when it is plotted as pairs of light:photosynthesis rate values (Fig. 3.8b). One domain of the sequence, in red, is consistently lower than the rest; i.e. for the same value of light the lower sequence is ~70% of the photosynthetic rate of the upper. This section falls between 40 and 70 minutes, marked by the dashed vertical lines on Fig. 3.8a, and it occurs when the local maxima of light flux are less relative to those occurring in the periods 1–40 and 71–120 minutes.

Estimates of A/Q curves for the two domains found that the greater photosynthesis, blue dashed line Fig. 3.8b, has values: A_{max} 9.1 CO_2 m^{-2} s^{-1}; θ 0.15; ϕ 0.13 $(h\nu_{PAR})^{-1}$; R_d 0.9 μmol CO_2 m^{-2} s^{-1}; and the dashed line for the lower value domain: A_{max} 5.9 CO_2 m^{-2} s^{-1}; θ 0.72; ϕ 0.074 $(h\nu_{PAR})^{-1}$; R_d 0.7 μmol CO_2 m^{-2} s^{-1}.

Clearly, changes in rate occur with change in light within both the periods of greater and lesser light flux; i.e. they follow a curvilinear response to changing light. The shift between domains can be attributed to overall differences in light flux, particularly decrease in local maxima, that may have produced, first, a decline in rca activity around 40 minutes and which would lead to reduction in active RuBisCo sites, and then an increase in rca activity around 71 minutes when light, and so ATP

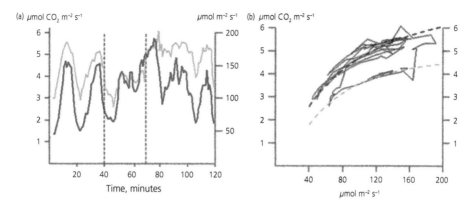

Fig. 3.8 (a) PAR (red) and parallel measurements of photosynthetic rate (green) with a pattern measured under forest conditions for *A. amabilis*. (b) Data from (a) with PAR as abscissa and photosynthetic rate as ordinate: black lines are for the periods 1–39 and 72–120 minutes, red values are for the period 40–71 minutes. Photosynthesis curves (equation 1) are fitted to the two periods and overlie the data as dashed lines.

and NADPH, increased. Salvucci et al. (1985) and Robinson and Portis (1989) suggest that rca is regulated by the levels of ADP/ATP and adjusts the rate of CO_2 assimilation to prevailing light flux, but of course NADPH, and possibly other reduced molecules, would be produced and drive essential reactions of the dark cycle.

There is continuous CO_2 uptake (Fig. 3.8) while light flux varies between 50 and 200 μmol m^{-2} s^{-1} in conditions where there were no sunflecks. In their review of sunflecks and their effects Way and Pearcy (2012) suggest the importance of sunflecks for carbon uptake in tropical ground flora species—but the contributions are variable: clearly sunflecks can result in photoinhibition, as discussed earlier in the chapter (section 3.4.1.2), Young and Smith (1983) describe the positive effects of diffuse light, relative to direct light on clear days, to an understory species. Diffuse light has contributions from the complete sky hemisphere and so provides more constant illumination. From eddy correlation measurements of CO_2 exchange on an old-growth *Nothofagus* forest Hollinger et al. (1994) report that greatest CO_2 uptake was on days of high proportion of diffuse light—even though they were days of lowest total light received.

3.6 Discussion

It has long been known that unshaded foliage developed under light flux typical of open conditions, frequently termed 'sun' foliage, has physiological, anatomical, and morphological characters different to 'shade' foliage (Lambers et al. 2008). For 'sun' foliage, A/Q curves, on a foliage area basis, generally have greater A_{max}, light saturation point (LSP) and light compensation point (LCP) than foliage developed under lesser light flux, 'shade' foliage. These differences in photosynthesis correlate with sun foliage being thicker but smaller in area with more layers of cells. Differences between 'sun' and 'shade' foliage have been defined for the shade-tolerant species *A. amabilis* by Tucker et al. (1987), Brooks et al. (1994, 1996), King (1997), and Mori et al. (2008) who also cite references to other species.

However, under field conditions foliage within an established canopy receives a time series of variable radiation in both upper and lower canopy. In the upper canopy light flux received by a foliage surface from direct light depends upon foliage angle, which is typically more upright than in the lower canopy, and foliage clumping, particularly for conifers (Niinemets 2006). Even during a day of constant sunshine the received light flux at a foliage surface will vary. In the lower canopy the duration of light flux of direct light depends upon interaction between gap size in the canopy and sun angle. A number of tissue and cellular protection and repair systems may function in concert with a continually changing rate of CO_2 assimilation. Their functioning depends upon the general conditions under which foliage developed and

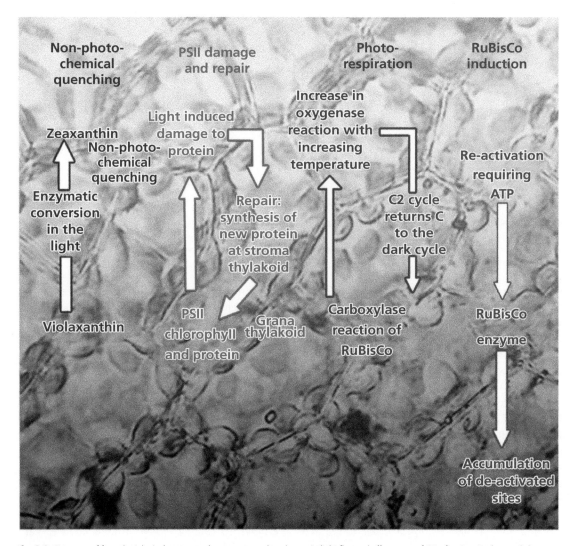

Fig. 3.9 Summary of four physiological processes that can respond to change in light flux and affect rates of CO_2 fixation. Background shows chloroplasts aligned along walls of leaf cells: iStock ID 1287371633 credit Videologia.

evidence to date suggests species differences in functioning.

Protection and repair systems (Fig. 3.9) operate in different ways. Violaxanthin reduces the excitation state of chlorophyll and so operates directly to reduce the flux of light to the photochemical process. Damage and repair of the PSII reaction centre directly modulates production of excited electrons. The C_2 photorespiration system uses O_2, reduces loss of fixed carbon, and maintains ATP-ADP and NADP-NADPH$^+$ recycling, discussed in Chapter 6.

This process can be considered as maintaining electron flow through the light reactions at higher temperatures and reducing possible ROS production. The rca induction of RuBisCo can be considered as a protection and repair system whereby active blocking of RuBisCo reaction sites in dark conditions maintains precursors of reactions in the dark cycle—continuous loss of function of active sites of RuBisCo is balanced by the ATP requiring action of rca, the RuBisCo chaperone, and provides a control system of counteracting processes between darker and lighter conditions. These controls are discussed in Part II. Stomatal control contributes to reducing occurrence of low water potentials but reduces CO_2 availability.

The systems represented in Fig. 3.9 are each likely to have some direct effect on CO_2 fixation and are indicative of the dynamic nature of physiological processes in response to change in the environment. Other controls may also exist. A notable system is chloroplast movement (Kong et al. 2013) whereby chloroplasts move towards cell walls under greater light, and genetic control of this process has been identified. Foliage movements that follow or avoid the sun also exist in some species (e.g. Kutschera and Briggs 2016; Rakocevic et al. 2018).

3.7 References

Bellucco, V., Marras, S., Grimmond, C. S. B., Järvi, L., Sirca, C. and Spano, D. 2017. Modelling the biogenic CO_2 exchange in urban and non-urban ecosystems through the assessment of light-response curve parameters. *Agricultural and Forest Meteorology*, 236, 113–122.

Brooks, J. R., Hinckley, T. M. and Sprugel, D. G. 1994. Acclimation responses of mature *Abies amabilis* sun foliage to shading. *Oecologia*, 100, 316–324.

Brooks, J. R., Sprugel, D. G. and Hinckley, T. M. 1996. The effects of light acclimation during and after foliage expansion on photosynthesis of *Abies amabilis* foliage within the canopy. *Oecologia*, 107, 21–32.

Buckley, T. N. 2005. The control of stomata by water balance. *New Phytologist*, 168, 275–291.

Carmo-Silva, A. E. and Salvucci, M. E. 2013. The regulatory properties of Rubisco activase differ among species and affect photosynthetic induction during light transitions. *Plant Physiology*, 161, 1645–1655.

Chazdon, R. L. and Pearcy, R. W. 1986. Photosynthetic responses to light variation in rainforest species. I. Induction under constant and fluctuating light conditions. *Oecologia*, 69, 517–523.

Chazdon, R. L. and Pearcy, R. W. 1991. The importance of sunflecks for forest understory plants. *BioScience*, 41, 760–766.

Englund, S. R., O'Brien, J. J. and Clark, D. B. 2000. Evaluation of digital and film hemispherical photography and spherical densiometry for measuring forest light environments. *Canadian Journal of Forest Research*, 30, 1999–2005.

Ernstsen, J., Woodrow, I. and Mott, K. 1997. Responses of Rubisco activation and deactivation rates to variations in growth-light conditions. *Photosynthesis Research*, 52, 117–125.

Farquhar, G. D., von Caemmerer, S. and Berry, J. A. 2001. Models of photosynthesis. *Plant Physiology*, 125, 42–45.

Gates, D. M. 1980. *Biophysical Ecology*. New York, Springer.

Genty, B., Briantais, J.-M. and Baker, N. R. 1989. The relationship between the quantum yield of photosynthetic electron transport and quenching of chlorophyll fluorescence. *Biochimica et Biophysica Acta (BBA)—General Subjects*, 990, 87–92.

Graham, L. E. 1993. *Origin of Land Plants*. New York, Wiley.

Hollinger, D. Y., Kelliher, F. M., Byers, J. N., Hunt, J. E., McSeveny, T. M. and Weir, P. L. 1994. Carbon dioxide exchange between an undisturbed old-growth temperate forest and the atmosphere. *Ecology*, 75, 134–150.

King, D. A. 1997. Branch growth and biomass allocation in *Abies amabilis* saplings in contrasting light environments. *Tree Physiology*, 17, 251–258.

Kitao, M., Lei, T. T., Koike, T., Tobita, H. and Maruyama, Y. 2000. Susceptibility to photoinhibition of three deciduous broadleaf tree species with different successional traits raised under various light regimes. *Plant, Cell and Environment*, 23, 81–89.

Kong, S.-G., Arai, Y., Suetsuga, N., Yanagida, T. and Wada, M. 2013. Rapid severing and motility of chloroplast-actin filaments are required for the chloroplast avoidance response in *Arabidopsis*. *Plant Cell*, 25, 572–590.

Kopp, G. and Lean, J. L. 2011. A new, lower value of total solar irradiance: evidence and climate significance. *Geophysical Research Letters*, 38.

Kutschera, U. and Briggs, W. R. 2016. Phototropic solar tracking in sunflower plants: an integrative perspective. *Annals of Botany*, 117, 1–8.

Kyei-Boahen, S., Astatkie, T., Lada, R., Gordon, R. and Caldwell, C. 2003. Gas exchange of carrot leaves in response to elevated CO_2 concentration. *Photosynthetica*, 41, 597–603.

Lambers, H., Chapin, F. S., III and Pons, T. L. 1998. *Plant Physiological Ecology*. New York, Springer.

Lambers, H., Chapin, F. S and Pons, T. L. 2008. *Plant Physiological Ecology*, Second Edition. New York, Springer.

Leakey, A. D. B., Scholes, J. D. and Press, M. C. 2005. Physiological and ecological significance of sunflecks for dipterocarp seedlings. *Journal of Experimental Botany*, 56, 469–482.

Ma, Z., Behling, S. and Ford, E. D. 2014. The contribution of dynamic changes in photosynthesis to shade tolerance of two conifer species. *Tree Physiology*, 34, 730–743.

Mori, A. S., Mizumachi, E. and Sprugel, D. G. 2008. Morphological acclimation to understory environments in *Abies amabilis*, a shade- and snow-tolerant conifer species of the Cascade Mountains, Washington, USA. *Tree Physiology*, 28, 815–824.

Niinemets, U. 2006. The controversy over traits conferring shade-tolerance in trees: ontogenetic changes revisited. *Journal of Ecology*, 94, 464–470.

R Development Core Team. 2013. *R: A Language and Environment for Statistical Computing*. Vienna, R Foundation for Statistical Computing.

Rakocevic, M., Müller, M., Matsunaga, F. T., et al. 2018. Daily heliotropic movements assist gas exchange and productive responses in DREB1A soybean plants under drought stress in the greenhouse. *Plant Journal*, 96, 801–814.

Robinson, S. P. and Portis, A. R. 1989. Adenosine triphosphate hydrolysis by purified Rubisco activase. *Archives of Biochemistry and Biophysics*, 268, 93–99.

Salvucci, M. E., Portis, Jr. A. R. and Ogren, W. L. 1985. A soluble chloroplast protein catalyzes ribulosebiphosphate carboxylase/oxygenase activation in vivo. Photosynthesis Research, 7, 193–201.

Seemann, J. R., Kirschbaum, M. U. F., Sharkey, T. D. and Pearcy, R. W. 1988. Regulation of ribulose-1, 5-bisphosphate carboxylase activity in *Alocasia macrorrhiza* in response to step changes in irradiance. *Plant Physiology*, 88, 148–152.

Slot, M., Wirth, C., Schumacher, J., et al. 2005. Regeneration patterns in boreal Scots pine glades linked to cold-induced photoinhibition. *Tree Physiology*, 25, 1139–1150.

Stanhill, G. and Cohen, S. 2001. Global dimming: a review of the evidence for a widespread and significant reduction in global radiation with discussion of its probable causes and possible agricultural consequences. *Agricultural and Forest Meteorology*, 107, 255–278.

Stenberg, P., Smolander, H., Sprugel, D. and Smolander, S. 1998. Shoot structure, light interception, and distribution of nitrogen in an *Abies amabilis* canopy. *Tree Physiology*, 18, 759–767.

Taiz, L., Zeiger, E., Møller, I. M. and Murphy, A. 2022. *Plant Physiology and Development*, Seventh Edition. Oxford, Oxford University Press.

Thornley, J. H. M. 2002. Instantaneous canopy photosynthesis: analytical expressions for sun and shade leaves based on exponential light decay down the canopy and an acclimated non-rectangular hyperbola for leaf photosynthesis. *Annals of Botany*, 89, 451–458.

Tucker, G. F., Hinckley, T. M., Leverenz, J. W. and Jiang, S.-M. 1987. Adjustments of foliar morphology in the acclimation of understory Pacific silver fir following clearcutting. *Forest Ecology and Management*, 21, 249–268.

van de Hulst, H. C. 1979. *Multiple Light Scattering: Tables, Formulas, and Applications*. New York, Academic Press.

Vico, G., Manzoni, S., Palmroth, S. and Katul, G. 2011. Effects of stomatal delays on the economics of leaf gas exchange under intermittent light regimes. *New Phytologist*, 192, 640–652.

Vierling, L. A. and Wessman, C. A. 2000. Photosynthetically active radiation heterogeneity within a monodominant Congolese rain forest canopy. *Agricultural and Forest Meteorology*, 103, 265–278.

Watling, J. R., Robinson, S. A., Woodrow, I. E. and Osmond, C. B. 1997. Responses of rainforest understorey plants to excess light during sunflecks. *Functional Plant Biology*, 24, 17–25.

Way, D. A. and Pearcy, R. W. 2012. Sunflecks in trees and forests: from photosynthetic physiology to global change biology. *Tree Physiology*, 32, 1066–1081.

Woodrow, I. and Mott, K. 1989. Rate limitation of non-steady-state photosynthesis by ribulose-1, 5-bisphosphate carboxylase in spinach. *Functional Plant Biology*, 16, 487–500.

Young, D. R. and Smith, W. K. 1983. Effect of cloudcover on photosynthesis and transpiration in the subalpine understory species Arnica latifolia. *Ecology*, 64, 7.

Zavafer, A., Chow, W. S. and Cheah, M. H. 2015. The action spectrum of Photosystem II photoinactivation in visible light. *Journal of Photochemistry and Photobiology. B, Biology*, 152, Part B, 247–260.

PART II

Plant control systems

Introduction to Part II

Part I describes plant growth as the result of interaction between multiple processes, each of which can be affected by different components of the environment. But growth is not a random process, despite fluctuations in the environment, and it results in an organized structure frequently of a characteristic size depending on species. Plant growth is controlled.

The science of control engineering provides valuable definitions of types of control systems, what they do, how they function, and their important characteristics (Chapter 4). Engineering control systems can be analysed in terms of the *command* applied to the control system which produces a particular response; the *operand* that the command acts on; and the *output* which is the actual response from the control system. Examples are given for engineering examples and some plant systems with similar control structures, illustrating that there are different types of control systems in plants that can affect growth. A major difference is that in plants genetic control can change the composition and structure of the system itself even during growth.

The circadian clock (Chapter 5) initiates growth. It is a nuclear-controlled biological oscillator that acts in *feedforward* control to schedule plant development and physiological processes. The WUS-CLV3 system of the plant apex operates as a *regulator*, ensuring that while cell differentiation and organ growth take place pluripotent cells are maintained that can continue the process: the inherent regularity in plant form is determined by actively controlled patterns of accumulation and depletion of auxin. The *feedforward* scheduler and *regulator* of the spatial structure of the plant together comprise the control processes that propel the initiation and continuation of shoot and foliage growth.

Chapter 3 shows that protection and repair of physiological systems are important: in particular that we should not think of the photosynthesis system simply having the sole function of CO_2 fixation. Multiple controls indicate that CO_2 fixation must be achieved whilst accommodating changes in the light environment and their potential effects, e.g. Murchie et al. (2018). Control of the response to such changes is achieved through the structures and processes in the photosynthesis system that can protect against, and repair, damage as well as maintain the capacity to increase CO_2 fixation following periods of low light (Chapter 6). These include continual change in chloroplast structure enabling physiological processes, reduction in activation of chlorophyll, repair of damaged PSII complexes, and maintaining electron flow through the light reaction system whilst reducing the potential for ROS formation. These processors act as *governors*, tending to reduce what could otherwise be the damaging effects of high light flux. Controls of the CBB cycle act to *regulate* capacity to respond to increase in light and under some conditions contribute to maintaining a non-damaging chemical environment within tissues.

The controls outlined in Chapters 5 and 6 provide regularity and stability in plant growth and development. Because the environment is variable in many of its characteristics the plant control systems must provide a dynamic response to produce regularity and stability.

Reference

Murchie, E. H., Kefauver, S., Araus, J. L., et al. 2018. Measuring the dynamic photosynthome. *Annals of Botany*, 122, 207–220.

CHAPTER 4

Control processes of plant growth

4.1 Introduction

Multiple plant characters interact to affect growth. The increase of maize yield achieved through plant breeding, discussed in Chapter 2, was associated with changes in developmental, morphological, and physiological characters. The survival of bristlecone pine appears related to its capacity to regenerate foliage after damage, longevity of its foliage in a generally harsh environment, and perhaps its capacity for slow growth. We are faced with the need to understand systems that depend upon interactions between multiple characters. This requires an integration of how variation in different types of character affects components of growth. An integrated understanding can be achieved by delineating the sequence of component processes that produce growth and determining how they are controlled.

Engineers have established an extensive science of control systems (Bennett 1993) for which they have defined the components and classified different types of control (DiStefano et al. 1990; Leigh 2012) with the objective of defining how systems respond to change and what characteristics of a system determine the types of response that occur. Their approach has long been extended to other than engineering systems including examples from biology, e.g. Ashby (1956), Milsum (1966), and terms from control engineering are used in describing plant systems—researchers frequently use the term *feedback* to describe interactions between component parts of the plant growth system, although almost invariably with a broader meaning than it has in engineering.

This chapter introduces the types of control that processes involved in growth may have. There is variation in how control is achieved, the role that component controls play in growth, and how effective individual control processes are, or can be. Individual control processes are discussed in more detail in subsequent chapters with regard to interconnections between types of processes and this leads to a description of growth in terms of interacting control systems.

4.2 Control processes

Control systems have the basic structure of a *command*, an *operand*, and an *output*.

- The *command* is applied to a control system or subsystem and produces a specified response.
- Commands may operate through different *signal types*.
- Control systems perform some function(s) on an *operand(s)*.
- An *output* is an actual response obtained from the control system.

The characteristics of *commands* and their *signals*, *operands*, and *outputs* are frequently different between control systems.

A classic example of engineering control is that of room temperature through a thermostat and heater (Table 4.1). Temperature is measured electronically and compared through the thermostat circuit against a requirement. If the measured temperature is less than that required by a certain amount, an electrical *signal* to the heater, the *operand*, causes it to add heat to the room, which is the *output*. The components of the control system itself, that is the measurement and comparison process and the heater, are not changed.

The Dynamics of Plant Growth. E. David Ford, Oxford University Press. © E. David Ford (2023). DOI: 10.1093/oso/9780192867179.003.0004

Table 4.1 The basic components of *command*, its *signal type*, *operand*, and *output* for room heating under thermostat control and reiteration of long shoots in *Pinus longaeva*.

System	Command	Signal type	Operand	Output
Room temperature control	Temperature deficit in measured relative to required	Electrical	Heater	Heat that increases room temperature
Long shoot reiteration in *P. longaeva*	Damage to apex of a long shoot	Involvement of a phytohormone	Buds of short shoots	Change of bud condition from quiescent to active and growth into a long shoot

Some control systems in plants act to maintain conditions within a system, but frequently plant control systems are involved with synthesis of new material that can cause the system to change. The reiteration of branches and foliage in bristlecone pine changes the developing structure of the plant (Fig. 2.4). There the *command* is damage to the apex of a long shoot, the *operand* is the bud of one or more short shoots, and the *output* is a new long shoot on which there are short shoots, some of which have buds that may become new *operands* (Table 4.1). The *signal* that induces growth of the bud has not been defined but, from analogy with apical decapitation experiments that induce lateral bud activity, may be assumed to be related to phytohormone balances. Of course this response is dependent upon a prior control system that produced potentially active buds on short shoots and an important question is how some buds on short shoots respond and others do not.

In both examples in Table 4.1 details of how the control processes function are not defined. In practice it is the details of the control processes that determine how control functions and different types of control are found in different situations, as discussed throughout this chapter.

4.2.1 Passive and active control

There are two broad categories of control systems, passive and active, that have different characteristics and have examples in characters that represent indirect and direct control respectively.

Passive control systems contain a component, or components, that maintain the organization of a system within certain limits through the dissipation of mass or energy which is produced by the properties of the passive control component. They are dynamic in that they come into operation under some particular condition.

Active control systems contain a component or components that respond to a command to change the condition of the system through expenditure of energy.

An engineering example of passive control is the inclusion of dashpots in buildings potentially subject to shaking by earthquakes. A dashpot is a damper that resists motion due to viscous friction. The building, or some part of it likely to move, rests on dashpots that absorb energy so that building movement is partially arrested. Design of a complete system for a building would require calculation of the correct number of dashpots with sufficient energy-absorbing capacity that movement of the building would be restricted to within a safe tolerance.

Leaf curvature in maize is an example of a passive control system. Curvature affects the intensity at which energy is intercepted and the duration of high interception flux density. This can affect utilization of light energy in photosynthesis and the potential avoidance of high energy loads (discussed in Chapters 2 and 8), with more upright foliage in maize being more effective at high-density planting. However, the benefit of more horizontal foliage at low planting density also needs to be considered in the appraisal of foliage curvature. The control system is passive in that curvature does not change once foliage has formed, but it is dynamic in that its effects depend upon the radiation flux that varies continuously. The *command* is radiation flux density and the *operand* is the foliage angle and azimuth relative to radiation source. The *output* may have a number of components of photosynthetic and radiation balance function and/or competition effects. The change in foliage curvature as planting density changes involves active control through alteration in spectrum of light received (*command*) changing a balance of phytohormone (*signal type*) that affects tissue development of the growing leaf (*output*) (Chapter 7).

Further examples of passive control include structures such as leaf hairs, e.g. *Stachys byzantia* (lamb's ear), *Encelia farinosa*, *Pittosporum crassifolium*, that can develop to cover leaf surfaces and increase reflectance of radiation or that may increase the thickness of a leaf boundary layer which can be important in maintaining a low transpiration rate of xerophyte leaves in windy and/or dry conditions. A problem in interpreting the function of such structures is that they may influence a number of processes, but nevertheless plant hair provides passive control.

The genetics of a plant determine control systems in two ways termed by Ashby (1956) as indirect and direct. Indirect control is where genetic information has defined construction of a structure and/or physiology that can operate as a control process, as for example leaf curvature in maize that affects both the amount of radiation that may be intercepted and the duration of different flux densities received. Direct control is where the products of gene expression are integrated in the control processes. The circadian clock, described in Chapter 5, is one of the most important examples of direct control, with genes actively producing products that during each day control expression of other genes, resulting in a biochemical oscillator that affects many processes over the diurnal period.

Within active control there is further division into open-loop, feedforward, and closed-loop control systems.

4.2.2 Open-loop control

An open-loop control system is one in which there is a distinct control but its action is independent of the output.

A domestic garden watering system is an example of an open-loop control system (Fig. 4.1). The times when the watering system is required to switch on and then off are set on a clock which, through an electrical *signal*, opens a solenoid valve at those set times, enabling water to flow, and then, subsequently, closes the solenoid, stopping

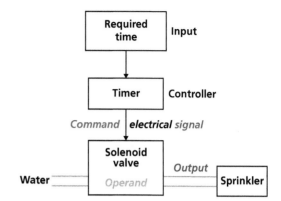

Fig. 4.1 Diagrammatic representation of a garden watering system as an open-loop control system. In this representation the required times for the system to come on and off, the *command*, are set manually in the timer. When a scheduled time is reached the controller operates a solenoid to switch the sprinkler on or off. There is no influence of environmental conditions or state of the soil on the operation.

the flow. The system is active because energy is required to switch on the solenoid valve and is open loop because operation of the clock (the *command*) is independent of any effect of application of the water (*output*)—particularly its effect on the soil—and operation is independent of any environmental perturbation. This system will switch on and then off regardless of soil conditions or weather.

The effectiveness of an open-loop control system depends upon its calibration; in the case of the garden watering system this would involve determining how often and for how long it should be switched on. Once calibrated the simplicity of open-loop control systems can make them very effective.

Fertilizer application to annual crop plants can be considered as an open-loop control system, at least under conditions where there is no monitoring of plant tissue during growth to check for additional fertilizer requirement. Fertilizer experiments can lead to recommendations by advisory services of how much fertilizer should be applied. For some crops, for which response to fertilizer seems variable, monitoring nutrient status can be used to adjust fertilizer application during growth, e.g. with canopy reflectance sensors for sugarcane (Amaral et al. 2018), so providing some closed-loop control.

The induction of photosynthesis when light flux density to foliage under low light is increased, discussed in Chapter 3, can be represented as an open-loop control system. The enzyme RuBisCo that performs the carboxylation reaction exists in conjunction with an activation enzyme, RuBisCo activase (rca), referred to as its chaperone (Carmo-Silva and Salvucci 2011). RuBisCo can become inoperative through molecular misalignment at the active sites and through night-time production of inhibitors that block the reaction sites. rca reactivates those sites through a process of molecular reconformation and is controlled by thioredoxin f, a product of the light reactions of photosynthesis (Chapter 6). The reduced form of thioredoxin is produced by the enzyme ferredoxin-thioredoxin reductase acting with reduced ferredoxin. (Thioredoxin activates a number of enzymes in the chloroplast stroma by reducing their disulphide bonds.) The action on RuBisCo also requires ATP (Fig. 4.2). In terms of control the *command* is increased light sufficient to produce thioredoxin f, the *signal*, and thioredoxin activates the enzyme. This signal defines the control as active. The *operand* is rca and the *output* is activated RuBisCo. This process is reviewed by Bhat et al. (2017).

This control of rca depends upon thioredoxin f but not RuBisCo, and in this sense the control is open loop. However, the rate of activation does depend upon the amount of available deactivated RuBisCo and this likely results in the exponential decrease in the rate of activation as measured by CO_2 uptake, seen in Fig. 3.7. And the dependence of activation on ATP can account for the slower rate of induction when a lower flux density of light is used as the *command*, as reported in Chapter 3. Of course the induction process is coupled with deactivation, but a number of conditions can result in deactivation of RuBisCo, e.g. high temperature and radiation flux density (Chapters 3 and 6), and active night-time production of inhibitors and open-loop control may provide a balance to them all.

4.2.3 Feedforward control

In some situations change in conditions that may affect a system can be predicted and where this occurs there can be feedforward control.

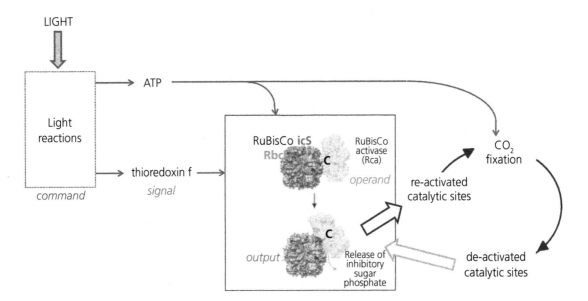

Fig. 4.2 Representation of photosynthesis induction as an open-loop control system with light as the *command*, rca as the *operand*, thioredoxin f as the *signal*, and reconfigured RuBisCo as the *output*. The enzymes RuBisCo (mainly green) and rca (blue) are represented in crystalline state. Enzyme structures reproduced from Bhat et al. (2016) with permission from Elsevier.

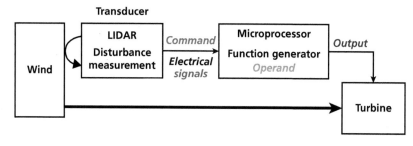

Fig. 4.3 Diagrammatic representation of feedforward control of a wind turbine. The LiDAR instrument faces into the wind and uses pulsed laser light to measure its speed at a known distance from the turbine blades. The *command* to the microprocessor results in a calculation of the pitch for the turbine blades that will provide steady speed and output from the turbine as the wind varies. Anticipatory control of this type can be valuable when the input to the process is variable and the time to change the functioning of the process, e.g. the pitch of the turbine blades, may be large relative to the timescale of variation in the wind.

In feedforward control the system responds to perturbing variation in a predefined way which is based on knowledge of the process.

An engineering example of feedforward control is anticipatory control of the pitch of the blades of a wind turbine (e.g. Schlipf et al. 2015) (Fig. 4.3). The objective of this control is to reduce variation in rotor speed and consequently structural loads on the tower supporting the turbine if the wind increases. The wind speed some distance away from the blades, but bearing down on them, is measured using light detection and ranging (LiDAR). The performance improvement depends on both the preview time in the wind speed measurement and establishing a relationship between the wind measurement and the wind that is actually experienced by the turbine. This relationship is defined in a microprocessor to make an adjustment to the pitch of the blades and so reduce variation in rotor speed. In contrast to feedback control, discussed next in this chapter, feedforward control acts the moment that variation in wind speed occurs without having to wait for a deviation in a process variable which would be required in feedback control. In engineering systems feedback control is frequently employed in conjunction with feedforward control because while feedback control systems can give accurate response to measured deviation they can be slow to respond.

The circadian clock, described in Chapter 5, is an essential feedforward control system for plant development and physiology. The perturbing variation is the diurnal pattern of night–day and how this changes seasonally and which influences development processes. The 'knowledge' of the process is contained in the linked sequence of genes producing products that may activate or suppress gene action and that can decay or be degraded with time.

4.2.4 Closed-loop control

In engineering closed-loop control is frequently designed to maintain constant output of a system. The control action is dependent on the output of the system matched to some required value and this has led to the concept of feedback.

A closed-loop control system is one in which the control action is somehow dependent on the output.

The four essential components of a closed-loop control system (DiStefano et al. 1990) are:

1. Measurement of the output that provides the primary feedback signal.
2. Calculation of the misalignment between feedback signal and reference input and translation of that to the controller.
3. A controller that uses the amount of misalignment to provide a signal to the process being controlled.
4. A component of the process being controlled that responds to the controller to adjust the output.

A thermostatically controlled heater regulating the temperature of a room is frequently used as an

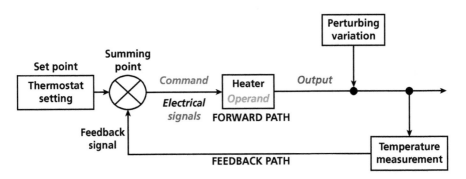

Fig. 4.4 Representation of room temperature control as a closed-loop feedback system. Room temperature is maintained within a tolerance to the set point. The summing point performs the calculation to produce a *command signal.* This can be designed to use different measures of the difference between the set point and measured temperature. Typically in domestic units the temperature measurement and summing point calculator are contained in the same wall-mounted unit along with the thermostat setting dial or switch.

example of a closed-loop control system (Fig. 4.4). The thermostat unit typically contains a dial or digital device for setting the temperature that provides a reference value (the set point) to the controlling circuits, a device such as a thermistor that measures and provides a signal of the actual room temperature to the controlling circuit. The command is determined by a component of the circuit (the summing point) that detects the difference between reference and measured values.

Typically closed-loop control systems are designed where there are two opposed time-varying influences on a system. In room temperature control the opposed influences are heating and natural cooling that can occur as a perturbing variation through changes in outside temperature or occasional door or window opening. When the thermostat detects that the measured temperature is less than the reference temperature (the set point) by a certain difference (tolerance) the heater provides heat until the measured temperature returns to be within less than the difference from the reference. The heater is then turned off until the measured room temperature again falls below the reference. The value of the difference is defined as part of the design of the thermostat.

The open-loop garden watering system could be changed so that a soil moisture measuring device is used to provide a feedback signal used to calculate a *command* that would replace a clock operating a switch. This would require defining a minimum soil moisture below which the system would be

switched on and a maximum soil moisture above which it would be switched off. It would also require placing the measuring device at a position to give a good measure of mean soil water conditions, which may not be a simple task. This would convert the garden watering system to a closed-loop control system. Closed-loop control of this general type is used in some commercial crop watering applications, e.g. Lea-Cox et al. (2013), where it can provide more effective use of water and a more even supply to plants over time.

From an engineering perspective *feedback* is the characteristic of a closed-loop control system that distinguishes it from open-loop systems. In the thermostat/heater system the command is equal to the input minus the output so that negative feedback exists in the system. The design of closed-loop control systems is an important task of control engineering (Leigh 2012). The word *feedback* came from work in the 1920s to achieve stability in the output from crystal oscillators being used in radio circuits—particularly to reduce frequency drift. Initially attempts were made to find open-loop circuits with stable components that could regulate frequency. The advance was to introduce negative feedback using measurements of frequency transmitted and frequency received. A benefit of negative feedback control is that a process that is not perfectly understood can nevertheless be controlled.

Closed-loop feedback control systems can provide accuracy, but this is only achieved with a certain tolerance within which conditions may oscillate. In design of feedback control systems care must be taken in specifying both the measurement of the *output* and calculation of the *command* signal from the set point and this measurement. In the example of the heated room if these produce a slow response to change then temperature may continue to fall after the required minimum temperature is actually reached, and then when heating is switched on the room temperature may overshoot. This and similar problems are well understood by engineers who design control circuits.

Three types of calculation can be made of the difference between set point and the feedback signal. Use of each in defining the *command* has advantages and disadvantages. These are typically referred to as P, I, and D.

- P: the difference in value (set point—feedback) is calculated and the command signal is proportional to this. The obvious advantage is that large differences can be reduced by a greater command signal. An important disadvantage is that correcting from large differences can result in an overshoot and possibly oscillations.
- I: an integration of past (set point—feedback) differences is calculated and the command signal is proportional to this. The advantage is that after a disturbance this type of control can return the output back to the exact set point—and because of this property integral controllers are sometimes called reset controllers. But a disadvantage is that the time over which an integration is made can result in a slow response and possibly a lack of stability.
- D: difference between successive (set point—feedback) values is calculated. The advantage of control based on this calculation is that it can improve the response to transient disturbance. However, it does not improve the error from steady state and can amplify the effects of random variation.

One standard type of command calculator evolved by engineers is the PID controller. This is designed to take into account these three types of deviation that can be calculated between required and measured values. A task for engineers is to estimate the value that each of these terms should contribute to the *command* signal, i.e. the gain of each, in particular applications.

Different characteristics of the environmental variation that may perturb the system may also affect design and operation of closed-loop control systems. In the example of room temperature control if the object is to achieve control over a wide possible range of exterior conditions, for example both summer and winter, then it may be necessary to add cooling, i.e. an additional component of the system that produces the output.

The term *feedback* is frequently used in describing interactions in the analysis of plant function. However, the meaning of the term varies between systems being described and is invariably different from the meaning of closed-loop feedback control defined by engineers. In biology the term *feedback* is generally used to distinguish between situations of simple linear progressive influence, where process A influences process B, from situations where there is reciprocating influence, where both A influences B and B influences A, whether directly or indirectly, and this is considered feedback. Ashby (1956) points out that while the concept of feedback is useful where a few components of a system interact, it becomes less useful where the number of components and interconnections between them becomes extensive. For understanding general principles of dynamic systems the concept of feedback as typically used in biology is insufficient and definition of the components of a control system and its characteristic behaviour is more appropriate.

4.3 Some analyses of control processes in plants

4.3.1 Homoeostasis and stability

An essential requirement for biological systems is that they continue to function effectively despite changes they experience in their environment. *Homoeostasis*, a frequently used concept, is:

a tendency of a biological system to resist change and to maintain itself in a state of stable equilibrium (Oxford English Dictionary).

The regulation of internal conditions in animals, particularly temperature in humans, is frequently given as an example of homoeostasis. While there are commonly accepted ranges for normal human temperature, e.g. 36.8 ± 0.4°C (oral measurement), in practice temperature ranges vary diurnally, annually, and with gender. There is a history of considering control of body temperature in terms of a negative feedback model, although a model of the type shown in Fig. 4.4 has proved difficult to apply, largely due to lack of finding how a set point could be established. Understanding this difficulty is instructive for consideration of control processes in biology.

Werner (2010) reviews various proposed definitions for set point in human thermoregulation and suggests: 'The long history of defence of the set-point concept in the thermoregulation literature seems to be mainly due to the inadequate adoption of this term from the technical concept of integral control to proportional control of body temperature.' In human temperature control there are known receptors (sensing processes) and effectors (processes that rectify deviations) and Werner notes that the effector mechanisms react proportionally to body temperature above/below a threshold and below/above a saturation. Proportional control, alone, cannot maintain a set point, but evidence for a set point and integral control seems absent.

Rather than considering thermoregulation as a process with a control structure similar to that in Fig. 4.4, both Kanosue et al. (2010) and Werner (2010) consider it as the result of interaction between the temperature control of two subsystems, the core body (organs, bones, muscle) temperature and the shell of the body (fat, skin). There are receptors and effectors associated with control of core and shell, both of which can be represented as open-loop systems. In the body core there is metabolic heat production, which can vary, and at the shell there is heat loss including through sweat rate. An important controlled link between the two subsystems is blood flow, controlled by vasoconstriction and vasodilation. Crucially the two subsystems interact in a closed-loop control system, with an output of one being input to the other and vice versa.

If temperature external to the body changes, interactions between the core and shell systems result in convergence to new temperatures for both core and shell, but the characteristics of the multiple receptors and effectors result in that new body temperature being within the accepted range. No comparison of signals is made within the system, as would be required for a system with a structure of the type shown in Fig. 4.4. The characteristics of receptors and effectors determine the response to temperature change and the progression of the core body temperature to a value within the necessary range. The system functions because of the evolution of receptors and effectors with finely tuned characteristics. Werner (2010) represents the shell control system as auxiliary to the core control system; Kanosue et al. (2010) represent it as a feedforward system—temperature changes are detected first in the shell. They note that receptors in the skin of both warm and cold are rate sensitive, suitable for feedforward control.

An important feature is that temperature control of the body as a whole is considered as the interaction between components of a distributed system. In engineering a *distributed control system* (DCS) is where control elements are distributed throughout the system with each component subsystem controlled by one or more controllers and these are linked to a central digital control computer so that control actions are local but there is central monitoring and scheduling.

4.3.2 Regulators

Regulators are a particular type of closed-loop control where the rate of a process is important rather than, say, a particular condition.

A *regulator* or *regulating system* is a closed-loop control system in which the reference input is generally constant for long periods of time, often for the entire time interval during which the system is operational. Such an input is often called a *set point*.

Regulators work to control the balance between two forces, such as in cruise control (speed control, autospeed, or tempomat) of a motor vehicle. In that case the forces that must be balanced are those of the motor against the combined resistances provided by the wheels on the road and vehicle body against the air. Regulators have to detect and respond to differences—detailed calculations of the component forces are not required, for example of the force that can be generated by the engine for any increase in throttle. In cruise control systems the set point can usually be changed manually to a lesser or greater speed. An essential quality of regulators is that they can function across a wide range of absolute values in the processes they control.

The production of foliage from the apical shoot meristem while it continues its own growth is described in Chapter 5 in terms of control through a regulator. Cells undergoing division occur in the centre of the meristem, whilst those capable of differentiation form a sheath around them. The phytohormone cytokinin stimulates cell division in the central cells, but this also results in production of a chemical that is actively transported to neighbouring cells where, in turn, it produces a repressor of cell division. The chemical and transport characteristics of this stimulator–repressor system maintain a balance between dividing and developing cells. There is no set point for this balance and it is affected by the supply of cytokinin that would stimulate growth as a whole.

4.3.3 Hysteresis

An important characteristic of some systems is that the effect of a control action is not constant but depends upon the system history. This is called *hysteresis*. Elastic hysteresis was one of the first types to be examined. If a rubber band is suspended vertically and a series of weights of increasing size are suspended from it the increase in length of the band can be measured. Then if the weights are removed in reverse order and length measurements repeated, the band will be found to have greater length for any particular weight. The band was harder to stretch when being loaded than unloaded. This is due to the molecular structure of rubber and how it changes when force is applied. Rubber regains its original structure but the molecular reconfiguration takes longer than the time taken to disrupt it. The change in cell volume in response to a decrease and then increase in water potential can show hysteresis (Chapter 9) depending upon the rates at which water potential change.

4.4 Some particular characteristics of control in plant systems

Although analogies with engineering control systems are useful for understanding some processes in plants and the roles they play in growth, there are important differences in the components of control and in the overall structure of some systems. Engineering systems typically operate in the domain of a single physical or chemical variable. For example, temperature control may involve both heating and cooling and may consider moisture in the air but can nevertheless be analysed in the domain of heat energy calculations. Most importantly the structure of the control system does not change during its operation. Even in detailed control systems, such as that for driverless vehicles, although multiple types of measurements may be made of speed and location, the control requirements can be calculated in the domain of motion and location. The control system itself does not change as a vehicle is underway.

In contrast, as a plant grows its structure changes and new components of control may come into play. The developmental change from vegetative to reproductive growth in maize (Chapter 2) is an example. Three features of plants involved in the control of growth need to be considered: structure, enzyme systems, and membrane controls.

4.4.1 Structure

Plant structure plays an important role in control of plant growth in two ways: process separation and meristem definition and limitation. Different processes must be located separately to function and there must be communication between locations. An obvious example is separation between the root and shoot. The transpiration stream and conduction of organic compounds require structures that enable these processes. An example of changes that take place in conduction tissues formed as plants age is given in Chapter 10.

Process separation can also be found at smaller scales. Within cells different functions take place in organelles. Within chloroplasts light causes damage to PSII which is embedded in thylakoid membranes of the chloroplast grana. Repair requires synthesis of a new protein which takes place in the stroma. PSII is reassembled in the grana margins and/or stroma lamella, following which the PSII returns to the stacked grana (Chapter 6).

The number and size of meristems may affect the amount of growth that can be made. This process is discussed in Chapter 7 where clonal differences in structure and physiology interact in affecting plant size.

Plant structure may be used to determine model structure where models are used in analysis, but there is no single unit that can be used to define the effects of either process separator or meristem limitations.

4.4.2 Enzyme systems

Most physiological processes in plants involve enzymes. Enzyme processes proceed at a rate determined by the concentration of the substrates it is operating on and this is typically curvilinear—the rate of product formed initially increases linearly as substrate concentration increases, but gradually a maximum value is reached when all active enzyme sites are operating at their maximum rates.

This curvilinear relationship for an enzyme with a single substrate, S, is modelled by the Michaelis–Menten equation which relates reaction rate, v, in the formation of the product, P, to the concentration of the substrate, [S]:

$$v = d\,[P]\,/dt = V_{\max}\frac{[S]}{K_M+[S]} \qquad (1)$$

Two parameters of this equation are used to characterize enzyme functions: maximum reaction rate ($V_{\max}$) and the substrate concentration, K_M, at which the rate of enzyme reaction is half that at $V_{\max}$.

For the same enzyme, parameter values can vary between related genotypes. Prins et al. (2016) investigated variation in RuBisCo between wild type and domesticated wheat relatives. The enzyme was extracted from young leaves, two to three weeks from sowing, and measurements of the enzyme function were made under controlled conditions so that the characteristics of the enzyme as both a carboxylase and oxygenase could be calculated. Measurements were made at two temperatures, 25 and 35°C.

For *Triticum aestivum* var. Cadenza, a widely grown bread wheat, where the substrate c indicates the carboxylation reaction and o the oxygenase, $v_c = 3.01$ μmol min^{-1} mg^{-1} (subscript, time, amount of enzyme) and $K_c = 16.3$ μM at 25°C, and 6.55 and 28.5 respectively at 35°C. Values for the oxygenase reaction at 25°C $v_o = 0.85$ μmol min^{-1} mg^{-1} and $K_o = 431.6$ μM, and 1.07 and 363.2 respectively at 35°C. Across the genotypes they measured at 25°C Prins et al. found ranges of $v_c = 1.74$–3.47 and $K_c = 11.9$–16.5. They calculated that genotypes with a higher v_c tended to have a lower affinity for CO_2. These figures may give some comparative information; however, Igamberdiev and Roussel (2012) analyse why the low ratio of substrate to enzyme for RuBisCo makes application of Michaelis–Menten kinetics inappropriate (Chapter 6).

4.4.3 Membrane controls

The uptake of ions by cells illustrates important features of active control at the cellular level. Plant cells have a wall composed mainly of cellulose, interior to which is the plasma membrane surrounding the cytosol and its contents. The cytosol in turn surrounds a vacuole and is separated from it by a membrane, the tonoplast. These membranes separate regions with different characteristics in a way that enables ionic movement between them to be controlled. The plasma membrane and tonoplast are composed of a double layer of phospholipid, with the hydrophobic region of each layer forming the centre of the membrane and hydrophilic regions on both outside and inside (Taiz et al. 2022, sections 1.6 and 8.4). Within the double phospholipid layer are embedded proteins that span the complete membrane and which are transporters of ions and molecules. The plasma membrane operates as both a boundary and selective conduction system between the extracellular space (apoplast) and cytosol. The tonoplast provides a similar function between the cytosol and vacuole. There is some direct connection through plasmodesmata between the cytosols of neighbouring cells.

Conditions within cells may vary depending upon cell activity. Typically, living cells have some function in the process of autogenic synthesis involved in cell growth and expansion and/or construction, storage, and transport of compounds. Consequently the cytosol is a site of some type of synthesis using ions and molecules typically obtained from the apoplast. The vacuole functions in water uptake and storage during cell expansion and as a location for storage of a range of metabolites including sugars and pigments.

Membranes regulate cytosol conditions. The plasma membrane and tonoplast maintain the balance of anions, cations, and water content within the cytosol as physiological requirements and conditions in the apoplast change.

The main force controlling transport of molecules and ions through membranes is the electrochemical potential which is determined by concentration and electrical potential gradients. The pH values outside the cytosol and within the vacuole are typically lower than within the cytosol and the electropotential differences are less in the cytosol. These differences are maintained by transfer of protons from the cytosol through the membranes *against* their concentration gradient; i.e. protons are moved from a lower to a higher concentration. This requires energy and is an active process. Across the plasma membrane this transfer is achieved through the enzyme plasma membrane H^+-ATPase (PM H^+-ATPase) and across the tonoplast by V-ATPase. Protons circulate back into the cytosol in association with transport of ions or molecules through secondary active transport proteins. Taiz et al. (2022, section 8.4) discuss the range of vacuolar acidification found in different species.

Unlike the ATPase embedded in chloroplast and mitochondria membranes, which operates as a molecular motor with a rotating subunit, PM H^+-ATPase is a linear polypeptide with 10 transmembrane segments and most of its structure is on the cytosolic side of the membrane (Falhof et al. 2016) including the N- and C-terminal domains (Fig. 4.5). In their review of the enzyme and its function Falhof et al. report that PM H^+-ATPase exists in 'an autoinhibited state, where ATP hydrolysis is only loosely coupled to H^+ transport, and an upregulated state with tight coupling between ATP hydrolysis and H^+ pumping'. Control of the enzyme's activity, and so of H^+ pumping, is by molecules interacting at different positions on the protein. For example, phosphorylation of the penultimate residue at the C terminal creates a binding site for proteins and this stabilizes the upregulated state of the enzyme by releasing C-terminal autoinhibition. Falhof et al. state: 'PM H^+-ATPase activity is regulated by virtually all factors known to regulate plant growth.' These include phytohormones and environmental conditions, so there are multiple possible signals. This one enzyme provides the conditions for multiple membrane transfer processes.

The requirement to maintain a pH gradient across the membranes involving expenditure of energy means that H^+ transport is an active process that is continuously available to maintain stable conditions for solute transport. Some molecules can diffuse directly through membranes, but there are three main classes of membrane transport proteins. These are channels and carriers that enable diffusion down the solute gradient of electrochemical potential and pumps that provide transport against the gradient (Taiz et al. 2022, Chapter 8). Channel proteins are membrane pores with their biophysical properties determining their transport specificity. For a given ion a membrane may have a variety of channels opening over different voltage ranges and providing specificity that enables ion transport to vary according to conditions. Carrier proteins do not have pores—the transported substance binds to the carrier protein, resulting in a conformational change in the protein which exposes the substance to the solution on the other side of the membrane. Carriers may transport 100–1,000 ions or molecules per second, while millions of ions can pass through an open channel. Active transport against the gradient of electrochemical potential is carried out by pumps using energy usually from ATP hydrolysis.

Aquaporins are proteins contained within membranes that are channels of water movement across the membrane. Water moves more quickly through aquaporins than through the membrane lipid bilayer and they can be changed between open and closed conditions.

With this system then control stability is maintained but not in terms of constant physical or chemical conditions but in the capacity to transport ions and molecules into and out of the cytosol as growing requirements and conditions in the apoplast change. Conditions can vary markedly with time,

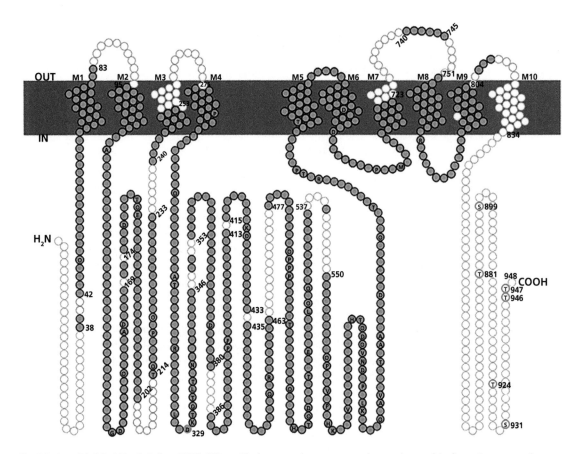

Fig. 4.5 A model of *Arabidopsis thaliana* PM H+-ATPase with plasma membrane represented across the top of the figure. Ten transmembrane segments (M1–M10) integrate the enzyme into the membrane. Most of the enzymes, including the N- and C-terminal domains, are on the cytosolic side of the membrane. Light-blue circles represent conserved regions between groups of plants in which inserts or deletions were never observed. Fully conserved amino acid residues are indicated with letters inside circles. White circles indicate regions where distantly related sequences do not align. Red circles mark in-vivo phosphorylated residues discussed by Falhof et al. Numbering of residues refers to the enzyme sequence. Reproduced from Falhof et al. (2016) with permission from Elsevier.

e.g. carbohydrate conditions in various tissues that vary seasonally and diurnally (e.g. Bansal and Germino 2009 for tree species in different environments; Atanosov et al. 2020 for accessions of *A. thaliana* from contrasting environments). There is a continuous process that adjusts conditions within the cytosol dependent upon the current function of the cell and conditions in the apoplast. Although the enzyme maintains a condition, i.e. the pH and electric potential difference across the membrane, this is not equivalent to homeostasis, which is limited deviation from some particular definable condition or concentration.

Membrane regulation of cytosol conditions is achieved locally at individual cells and there are no centrally defined set points or feedback as in a closed-loop control system. In general, PM H+-ATPase regulation can be considered as a conditional switch, i.e. triggered by removal of its autoinhibition but requiring energy. Its function and therefore the role it plays in control must be considered in its effect on transport through membranes and in relation to other transporters (Zhou et al. 2021). There are multiple *commands* that can either remove the autoinhibited state of the enzyme, such as auxin which stimulates cell wall growth due to the effects of release of acidity into the developing wall structure, or inhibit growth, such as abscissic acid: PM H^+-ATPase is the *operand* and the *output* is proton transfer across the plasma membrane and there is variation in the structure of PM H^+-ATPase.

4.5 Discussion

Plant growth control processes are determined by a plant's genetics. There are five groups of control systems that each have distinct genetically determined characteristics that determine how their functions are performed in relation to specific environmental characteristics. These groups are listed along with reference to more detailed discussion in subsequent chapters. How these groups interact in the integrated process of growth and some types of variation that are found is also indexed.

1. A *feedforward* control of plant development driven by the circadian clock initiates and maintains growth activity and schedules particular growth activity or events in relation to changes in the environment. This control is described in Chapter 5 and examples of the importance of development are given in Chapters 2 and 9.
2. A *regulator* ensures symmetry in the species pattern of foliage production. The basics of the regulator are discussed in Chapter 5 and examples of the effects of variation in morphology on growth in Chapters 7 and 8.

Two essential requirements for plant growth are, of course, light and water. In both cases control systems govern what can be considered as their use in producing growth, but there are also controls that operate when imbalance occurs.

3. The autotrophic process of CO_2 fixation might initially be considered an *open-loop* control. However, under high light the capture and utilization process can lead to excessive production of chemical reduction capacity and potentially to cell and tissue damage. There are a number of processes that *protect* against this, so that photosynthesis is best considered as under *closed-loop* control. This is discussed in Chapter 6.
4. Cell expansion is driven by hormonal loosening of the primary cell wall and water uptake into the cell to maintain cell turgor pressure that ensures expansion. A positive control system maintains turgor pressure but may fail if tissue water potential decreases to certain levels. This system is discussed in Chapters 9 and 10.
5. Developmental processes are regulated through the circadian clock; changes in the environment can result in changes in both the structure of organs produced (*plasticity*) (Chapter 8) and the characteristics of physiological processes (*acclimation*).

4.6 References

Amaral, L. R., Trevisan, R. G. and Molin, J. P. 2018. Canopy sensor placement for variable-rate nitrogen application in sugarcane fields. *Precision Agriculture*, 19, 147–160.

Ashby, W. R. 1956. *An Introduction to Cybernetics*. New York, John Wiley.

Atanasov, V., Furtauer, L. and Naegele, T. 2020. Indications for a central role of hexokinase activity in natural variation of heat acclimation in *Arabidopsis thaliana*. *Plants (Basel)*, 9, 819.

Bansal, S. and Germino, M. J. 2009. Temporal variation of nonstructural carbohydrates in montane conifers: similarities and differences among developmental stages, species and environmental conditions. *Tree Physiology*, 29, 559–568.

Bennett, S. 1993. *A History of Control Engineering 1930–1955*. Stevenage, Peter Peregrinus Ltd on behalf of the Institution of Electrical Engineers.

Bhat, J. Y., Miličić, G., Thieulin-Pardo, G., et al. 2017. Mechanism of enzyme repair by the AAA^+ chaperone Rubisco activase. *Molecular Cell*, 67, 1–13.

Carmo-Silva, A. E. and Salvucci, M. E. 2011. The activity of Rubisco's molecular chaperone, Rubisco activase, in leaf extracts. *Photosynthesis Research*, 108, 143–155.

DiStefano III, J. J., Stubberud, A. R. and Williams, I. J. 1990. *Feedback and Control Systems*, Second Edition. New York, McGraw-Hill.

Falhof, J., Pedersen, J. T., Fuglsang, A. T. and Palmgren, M. 2016. Plasma membrane H^+-ATPase regulation in the center of plant physiology. *Molecular Plant*, 9, 323–337.

Igamberdiev, A. U. and Roussel, M. R. 2012. Feedforward non-Michaelis–Menten mechanism for CO_2 uptake by Rubisco: contribution of carbonic anhydrases and photorespiration to optimization of photosynthetic carbon assimilation. *Biosystems*, 107, 158–166.

Kanosue, K., Crawshaw, L. I., Nagashima, K. and Yoda, T. 2010. Concepts to utilize in describing thermoregulation and neurophysiological evidence for how the system works. *European Journal of Applied Physiology*, 109, 5–11.

Lea-Cox, J., Bauerle, W., Van Iersel, M., et al. 2013. Advancing wireless sensor networks for irrigation management of ornamental crops: an overview. *HortTechnology*, 23, 717–724.

Leigh, J. R. 2012. *Control Theory—A Guided Tour*, Third Edition. London, Institution of Engineering and Technology.

Milsum, J. H. 1966. *Biological Control Systems Analysis*. New York, McGraw-Hill.

Prins, A., Orr, D. J., Andralojc, P. J., Reynolds, M. P., Carmo-Silva, E. and Parry, M. A. J. 2016. Rubisco catalytic properties of wild and domesticated relatives provide scope for improving wheat photosynthesis. *Journal of Experimental Botany*, 67, 1827–1838.

Schlipf, D., Haizmann, F., Cosack, N., Siebers, T. and Cheng, P. W. 2015. Detection of wind evolution and Lidar trajectory optimization for Lidar-assisted wind turbine control. *Meteorologische Zeitschrift*, 24, 565–579.

Taiz, L., Zeiger, E., Møller, I. M. and Murphy, A. 2022. *Plant Physiology and Development*, Seventh Edition. Oxford, Oxford University Press.

Werner, J. 2010. System properties, feedback control and effector coordination of human temperature regulation. *European Journal of Applied Physiology*, 109, 13–25.

Zhou, J. Y., Hao, D.-L. and Yang, G.-Z. 2021. Regulation of cytosolic pH: the contributions of plant plasma membrane H^+-ATPases and multiple transporters. *International Journal of Molecular Sciences*, 22, 12998.

CHAPTER 5

Processes producing organization in the plant

5.1 Introduction

Although there is great variation in growth form between plant species, individual plants within a species typically have a common and identifiable regularity in their form. This coherence is largely achieved through repetition of genetically controlled shoot and foliage development sequences but is not produced by some centralized locus in the plant. Lack of such a dominant central control distinguishes plants from animals that have a central nervous system and/or a circulatory vascular system. In plants the essential aspects of both growth and its control are themselves distributed.

Growth is initiated, and continued, by the developmental processes that schedule physiological processes and growth within diurnal and seasonal patterns of environmental change. Environmental influences operate through effects on the genetic networks that produce growth, but a function of these networks is to produce a regulated plant structure. A primary location of growth is at plant apices, which generate structural organization. The basic processes involved in gene expression are central to the control systems of both development and production of new tissue. Gene expression itself is controlled by a number of processes that provide different types of control. The circadian clock provides anticipatory control for responses to diurnal and seasonal variation in the environment. The apical meristem, and the tissues produced by it, controls the production and spatial organization of plant structure through a combination of regulator control of meristematic tissue and a balance of feedback components of cell growth in the initiation

of lateral organs such as leaves. Development processes determine when growth can take place and meristem organization determines what can be grown; together they enable growth that produces a continuously organized system.

5.2 The basics of nuclear control

The cell nucleus *actively* regulates physiological processes and responses to environmental changes, and produces growth with specific structures. Fig. 5.1 provides a summary of the nuclear control system.

There are three stages in the process from gene to realization of gene action (Fig. 5.1), each of which may involve control processes and can be involved in duration of gene action:

1. Transcriptional regulation that controls production of messenger RNA (mRNA) and which requires the action of a promoter.
2. Post-transcriptional regulation where mRNA and ribosomes are transported from the nucleus to the cytoplasm and constituted to synthesize a protein.
3. Post-translational regulation—particularly of stability and longevity of the protein that has been produced.

Each of these stages can be controlled by genetic processes. It is through these controls that active regulation of physiological and growth processes is achieved. Greater detail of the process can be found in Chapter 3 of Taiz et al. (2022).

In stage 1 (Fig. 5.1) the enzyme RNA polymerase makes an mRNA transcript complementary to the gene DNA sequence. To initiate action the RNA

The Dynamics of Plant Growth. E. David Ford, Oxford University Press. © E. David Ford (2023). DOI: 10.1093/oso/9780192867179.003.0005

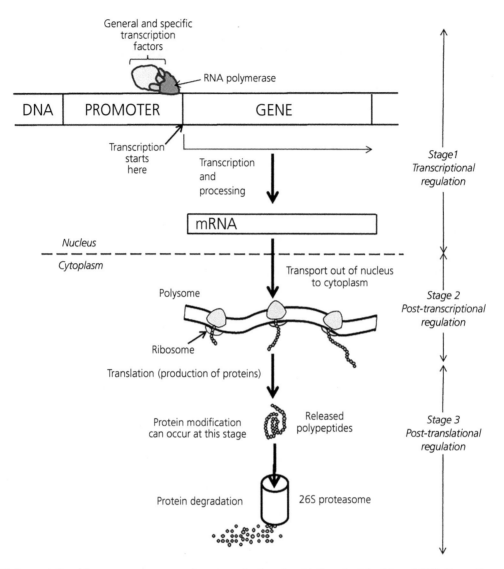

Fig. 5.1 Representation of three component processes of gene expression through protein formation (after Taiz *et al*. 2018, Chapter 2). Environmental influences can affect promoters of gene transcription, either positively or negatively, and controls can also occur on the transport or action of mRNA (post-transcriptional regulation) and on the proteins formed and which may provide the control function (post-translational regulation). The duration of gene action can be affected by the balance between rate of protein formation and protein degradation. Reproduced from Taiz and Zeiger (2010) *Plant Physiology* 5th Edition, Chapter 2, with permission from Oxford University Press.

polymerase requires general transcription factors, proteins positioning it on the DNA, and most transcription requires specific factors that bind to the DNA sequence upstream of the gene and so become part of transcription initiation. There can also be distal regulatory sequences on the DNA that are activators, repressors, or enhancers of transcription.

The most distally located transcription factors may interact with the initiation process through bringing them into contact with the promoter region by looping of the DNA. Regulation of transcription can occur through movement of protein into the nucleus after it has been synthesized in the cytoplasm to affect these transcription factors.

In stage 2 the mRNA is processed for transport to the cytoplasm and use for production of the protein. mRNA coming from chromosome DNA contains sections not functional in protein synthesis and corresponding to DNA introns—sections of DNA that do not code for protein and interrupt the gene sequence. These are removed from the mRNA which is spliced into a complete and functional strand. Stability of mRNA, and so its longevity and duration of activity, can be affected by binding proteins that can either stabilize it or make it more vulnerable to degradation by nucleases. mRNA stability can also be regulated by several types of small RNA molecules that do not code for proteins but are components of the RNA interference pathway (RNAi). RNAi is a process in which RNA molecules inhibit gene expression or translation by neutralizing targeted mRNA. Two types of small RNA molecules can be involved in this process: microRNAs, which arise from RNA polymerase transcription of particular sections of DNA, and small interfering RNA (siRNA).

In stage 3 proteins produced by mRNA both function and are degraded. The rate of degradation obviously affects duration of action. Degradation is an active process involving attachment of a small peptide, ubiquitin, via a ubiquitin ligase, to the protein. This ubiquination is an energy-requiring process. The protein can then be degraded by the 26S protease. There are many protein-specific ubiquitin ligases that regulate the turnover of specific proteins.

5.3 Coordinated scheduling through the circadian clock

The night and day cycle is a dominant feature of environmental variation experienced by plants. The circadian clock (Taiz et al. 2022, their section 20.3) is a nuclear control system providing anticipatory control for these daily changes. It has two functions:

1. To initiate expression of genes involved in cell function and growth at particular times in the diurnal cycle—referred to as *gating*. The circadian clock acts as a dynamic scheduler for the processes involved in plant growth which requires synthesis of components from different subsystems. Harmer et al. (2000) found that 6% of more than 8,000 genes examined show circadian changes in steady-state messenger array levels. Farré and Weise (2012) review evidence for circadian regulation of pathways in primary metabolism.

2. To account for seasonal variation in time of day-to-night and night-to-day change by the process of entrainment. These changes can be considerable. At mid-latitude of the temperate zone (45° from the equator) the time between sunrise and sunset at the summer solstice is ~15 h 36 minutes. Including the periods of civil twilight, when objects remain visible (sun no greater than 6° below the horizon) extends the light period to 16 h 21 minutes. The reciprocal shortening of the light period at the winter solstice means that plants experience a substantial range of light periods. The range of light periods between summer and winter is greater at higher latitudes.

The circadian clock is a genetically driven biochemical oscillator (Novák and Tyson 2008) with action initiated in the nucleus. A number of genes are involved in regulation, but this chapter initially considers the basic control action of a system with just one gene (Fig. 5.2). Transcription of this gene, controlled by the promoter region of its DNA, produces an mRNA which passes from the nucleus to the cytoplasm where it functions to synthesize a protein. Consider that this protein may inhibit further transcription of the gene through an effect on transcription factors, but also that the protein may be broken down enzymatically by proteolysis (stage 3, Fig. 5.1), and that mRNA degrades over time (Fig. 5.2a).

Consequently concentration of protein produced by the gene will first increase, but with a time delay due to the time taken for the processes of transcription, mRNA transport, and protein synthesis (A and B in Fig. 5.2a). The protein will transport to the nucleus (C in Fig. 5.2a) and inhibit gene transcription. Proteolysis of the protein (D and E in Fig. 5.2a) will decrease protein concentration, so enabling transcription again. The concentration of the protein will oscillate with period and magnitude depending upon the relative rates of protein synthesis, transport, and proteolysis. In the

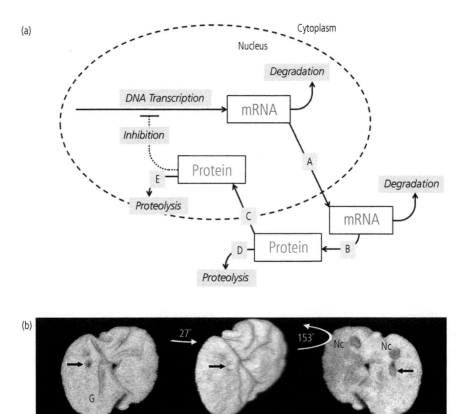

Fig. 5.2 (a) Basic structure of a nuclear-driven, single-gene, biochemical oscillator based on protein formation and inhibition of transcription by that protein as descried in the text (after Novák and Tyson 2008.) Processes at points A through E are described in the text. (b) Three-dimensional reconstruction of the surface of onion epidermal nuclei based on optical sections at 2.0-μm intervals from the surface. Rotation through 27° shows grooves, as labelled in the leftmost representation (G), across the surface and invaginations (marked by black arrows). In the further 153° rotation, Nc indicates nucleoli; scale bar = 10 μm. Part (b) reproduced from Collings D. A. et al. (2000) with permission from Oxford University Press.

simplified system in Fig. 5.2 the gene is the *operand*, the *command* is the promoter of transcription, and the *output* is the protein that has subsequent effects.

Although the plant cell nucleus is frequently represented as having a rounded structure, it can have grooves and invaginations with cytoplasmic cores that are stable structures (Fig. 5.2b). This increases nuclear surface area and Collings et al. (2000) report their 'apparent preference for association with nucleoli, and the presence in them of actin bundles that support vesicle motility suggest that the structures might function both in mRNA export from the nucleus and in protein import from the cytoplasm to the nucleus'.

Novák and Tyson define four essential requirements for biochemical oscillators:

1. There must be an inhibiting action between components: in the example just given it is self-inhibition but in circadian clocks there are genes that repress the actions of others, and genes that self-repress, as well as genes that activate other genes.
2. There must be delay in the action which is provided by the time required for synthesis and transport of mRNA and protein.
3. The component reactions must be non-linear, which is a characteristic of enzymatic processes.
4. Production and proteolysis must occur at time intervals that produce oscillation.

When there are antagonistic components in a process, in this case activation and inhibition, then oscillation is generally a possibility. In the case of the thermostatically controlled room temperature system (Chapter 4), where a constant temperature is required, the control system must be designed to protect against oscillation. In the circadian clock, oscillation is required in the output and that it should occur with a particular frequency and amplitude,

Shim and Imaizumi (2015) present a synthesis from multiple reports of protein expression through the day of component genes of the circadian clock in *Arabidopsis thaliana* (Fig. 5.3). The essential feature is a succession of maxima in expression levels for genes over the day.

This cycle is self-propagating; that is, if plants are grown in a 12-h light–12-h dark cycle and then, after a few cycles, the light period is extended without a dark interruption the expression cycle nevertheless continues. In this condition, referred to as 'free running', and when the cycle is not entrained its period is typically longer than 24 h.

Much research into the circadian clock conducted with *Arabidopsis*, which has the advantage of a defined genome, has produced a catalogue of many mutants with defined genetic changes and a plant that can be grown rapidly for experimentation. Comparative studies with other species have shown considerable conservation of key circadian clock genes, i.e. that they are found in other species (e.g. Khan et al. 2010; Filichkin et al. 2011). However,

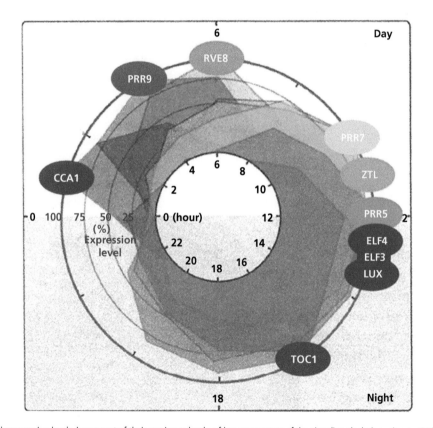

Fig. 5.3 Protein expression levels, in per cent of their maximum levels, of key components of the circadian clock throughout a 24-h day. By convention the clock is considered to start at the beginning of the light period, labelled 0 hour, and is counted on a 24-h basis. Gene names, with their abbreviations shown in white capitals, are listed in the text. Reproduced from Shim and Imaizumi (2015) with permission from Takato Imaizumi, University of Washington.

there are likely some differences between species. Gyllenstrand et al. (2014) found diurnal cycling of circadian clock genes in needles of Norway spruce (*Picea abies*) seedlings, but the clock consisted of fewer components than the angiosperm clock. The cycling dampened more rapidly and they concluded that there may be some differences in clock function between gymnosperms and other plant taxa. The number of operational clock genes may vary both between tissues and between species.

Gene names, which typically refer to a gene's first known action, are written in capital letters and are referred to by standardized abbreviations. The morning *CIRCADIAN CLOCK ASSOCIATED1 (CCA1)* and *LATE ELONGATED HYPERCOTYL (LHY)*, not shown in Fig. 5.3 as an independent protein but occur with CCA1 as a heterodimer, act together to activate the expression of *PSEUDO-RESPONSE REGULATOR 9 (PRR9)* and *PRR7* and to suppress expression of evening genes. They also suppress their own expression (Shim and Imaizumi 2015). In the early afternoon *PPR9* and *PPR7* suppress the expression of *CCA1* and *LHY*. Shim and Imaizumi provide a comprehensive description of suppressor and activator interactions through the cycle.

Two pathways have been suggested that entrain the circadian clock to the current light–dark periodicity. The first is through direct sensors of light: either the *ZEITLUPE/ GIGANTEA (ZTL/GI)* complex and its effects on *PRR5* and *TIMING OF CAB EXPRESSION 1 (TOC1)* (Shim and Imaizumi 2015) or the plant hormone phytochrome acting on a number of genes (Inoue et al. 2017). This action is a direct property of the circadian genes. The second is through the effect of sucrose as the hours of light change, providing a feedback signal *from* physiological processes of the cell *to* the clock that changes expression of evening genes through post-translational processes.

ZTL, which has a maximum protein expression late in the day (Fig. 5.3), is a blue light photoreceptor. Its abundance is regulated by a blue-light-dependent interaction with the evening phase large protein *GI* which protects it from proteasome degradation. *ZTL* degrades *PRR5* and *TOC1* in the dark. The *ZTL/GI* complex is a light-dependent, post-translational regulator that controls the period and amplitude of circadian clock gene expression (Shim and Imaizumi 2015).

Haydon et al. (2013) demonstrate that rhythmic sugar production from photosynthesis can entrain the circadian clock to the current light period. Sucrose activates *CCA1* through repression of PRR7 transcript abundance, which normally peaks in the late afternoon, resulting in extension of *CCA1* action. Starch production and utilization for night-time growth in *Arabidopsis* is closely controlled by the circadian clock (e.g. Graf and Smith 2011).

The circadian clock has a controlling effect on processes associated with photosynthesis and carbon metabolism. Catalase genes involved in regulation of ROS, the product of the light reactions of photosynthesis, and which are potentially damaging to tissue (see also Chapter 6), are controlled by the circadian clock (reviewed by Greenham and McClung 2015; Shim and Imaizumi 2015). Seki et al. (2017) propose that starch turnover is controlled by the circadian clock through sucrose signals. In *Arabidopsis* the correct timing of starch metabolism by the circadian oscillator is required for optimization of growth (Webb and Satake 2015).

The circadian clock may vary between tissues within the plant and there is indication of communication between tissues in setting timing sequences. Para et al. (2007) identify differences in temporal expression of clock components between vascular and the rest of leaf tissues in *Arabidopsis* and found that *PRR3* regulates the levels of the evening-expressed *TOC1* and that its abundance is increased in the vascular tissue. The expression profiles of clock genes in the vascular tissue (evening maxima) are inverse to that in the mesophyll (morning maxima) (Endo et al. 2014). The vascular clock has distinct regulatory targets relative to the mesophyll and a more persistent circadian rhythm under free-running conditions. Through experimentally perturbing the clock in either the mesophyll or vascular tissue Endo et al. found perturbation of the vascular clock affected *TOC1* expression in both mesophyll and vascular tissue but perturbation of the mesophyll clock only affected the mesophyll. They suggest the leaf vascular clock is distinct and robust and is able to control gene expression in neighbouring mesophyll cells.

There is indication of shoot→root interaction, so providing some indication of whole plant integration of the *Arabidopsis* circadian system (Takahashi et al. 2015). James et al. (2008), also investigating *Arabidopsis* and some of its mutants, found that in roots the morning genes of the clock (*CCA1* and *LHY*) operate but the mid-day and evening genes are decoupled: *TOC1* does not contribute to the period of the root clock in mature *Arabidopsis* plants. Under long day (LD) conditions (more than 12 h light in the 24-h period) the clocks in roots and shoots are synchronized and this synchronization depends upon photosynthesis. Application of sucrose at dusk in LD delayed and extended the next expression of *CCA1* and *PRR9* in roots but not shoots.

Novák and Tyson (2008) refer to the inhibition process that occurs between component genes of the clock as negative feedback in the sense, as they say, that it 'is necessary to carry a reaction network back to the starting point of its oscillation'. This is a different meaning to that defined by closed-loop feedback control in engineering (Chapter 4). In operation the progression of the circadian oscillator sequence is through a linked series of forward steps. In operation of the clock itself there is no sensing of its own function to condition expression of clock genes—so there is no closed-loop feedback control.

Although entrainment through the photosynthesis→sucrose effect is represented as a closed-loop control system (Fig. 5.4), this differs from the classic engineering example of feedback. The entrainment adjustment is not based on a calculation of the difference between actual length of the light period and that achieved by the clock but on the specific characteristics of gene action, protein effects on transcription, and proteolysis characteristics of protein, and degradation of mRNA. The balance of day gene to night gene expression is disrupted to increase that of day genes, whether through action of blue light→ZTL or photosynthesis→PRR7. In effect the clock operates as a *feedforward controller* but one that provides regular change in its control of physiological processes. Light-based control signals adjust the timing of the change, although this adjustment is small relative to the dominant cycle provided by the core clock genes. The control system represented in Fig. 5.4 does not include interactions with other cells and tissues, but this is clearly an important and developing subject.

5.4 Production of spatially regular organs by the shoot apical meristem

Foliage and lateral shoots may be arranged along a parent shoot in various ways, for example, alternately, in opposite pairs, or spirally. The variety of arrangements is illustrated by Bell (2008) who codifies what he terms the constructional organization of plants; the essential feature is an inherent regularity.

Control processes of shoot growth and production of laterals have been established for *Arabidopsis*, greatly aided by the identification and analysis of mutants with abnormal growth and development (Taiz et al. 2022, section 22.3). Three components are necessary to produce organized structure: (i) the shoot apical meristem (SAM) must continue extending, (ii) while producing new organs at regular intervals, and (iii) with a definable structure and size.

5.4.1 Maintenance of apical meristem stability

The SAM (Fig. 5.5) has a growth process that maintains a specific size. The domed multilayered structure with a cytohistochemical zonation is common to ferns, gymnosperms, and angiosperms, although meristem size varies between species. Continuing growth requires that when cells at the apex central zone (CZ; Fig. 5.5b) divide, some daughter cells must maintain the CZ and remain undifferentiated, while others move to the peripheral zones (PZs) and become involved in production of lateral organs.

Variability in growth rates between sectors is an important part in determining the geometry of the apical meristem (Truskina and Vernoux 2018). Cells in the CZ (Fig. 5.5b) divide more slowly than those in the PZ, with division cycles for *Arabidopsis* of 36–72 h in the CZ compared to 18–48 h in the PZ (Gaillochet and Lohmann 2015). The CZ has greater stiffness than both peripheral regions and the lateral organ primordia. In the L1 layer new division planes orient along maxima of mechanical tensions in the cell (Louveaux et al. 2016). Some daughter

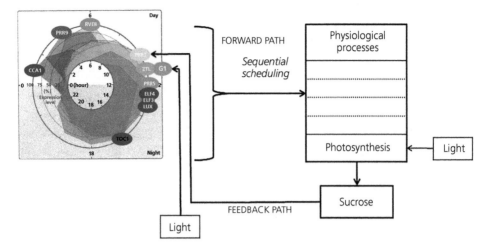

Fig. 5.4 Interaction between the circadian clock acting as a feedforward scheduler of physiological processes and sucrose from photosynthesis acting to entrain the clock to the current light–dark periodicity. Light acting directly at the *ZTL/GI* complex can also entrain the clock.

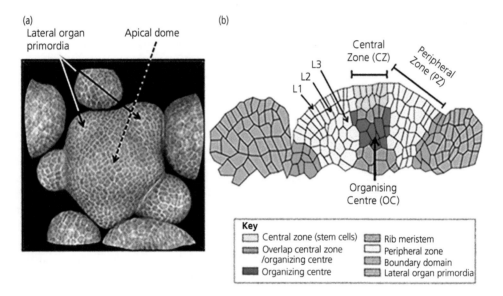

Fig. 5.5 (a) View of the apical meristem of *A. thaliana* from above, 25 days after germination and at the inflorescence stage showing lateral organ primordia and surrounding lateral organs. (b) Schematic of tissue organization in a longitudinal section of the meristem. The central dome-shaped structure has two outermost layers, L1 with L2 below it. Division in these layers is anticlinal whilst division in L3 is in all direction. Reproduced from Gaillochet and Lohmann (2015) with permission from The Company of Biologists under a CC BY 3.0 license.

cells from the CZ are displaced towards the PZ. Production of new organs at regular intervals requires that groups of cells divide actively and form lateral organ primordia, but such groups are separated by cells that do not divide marked as in the boundary domains (Fig. 5.5b).

The phytohormones cytokinin and auxin are both involved in cell growth and development of the SAM tissue (Taiz et al. 2022, section 18.3). Generally, cytokinin promotes cell division while auxin produces cell expansion—but their activities depend upon the balance between them at various points in the tissue. Both phytohormones can be considered as messenger molecules that act in gene transcription. Auxin can be transported actively from cell to cell via cell membrane bound efflux carriers called PIN proteins that are located at the cell poles so that transport is directional. (PIN proteins are integral membrane proteins involved in auxin transport. They get their name from the *Arabidopsis* pin1 mutant needle-like inflorescence phenotype, devoid of stem and leaves.) Control of cell activities in different zones of the meristem involves both activation and repression of gene activity as well as accumulation of contrasting regions of high and low concentration, particularly of auxin.

Cells in the CZ are considered to be stem cells: that is cells that are pluripotent—able to give rise to all cell types (see Gaillochet and Lohmann 2015 for a discussion on this definition particularly in comparison to its use in animal biology). The number of stem cells in the CZ is maintained but balanced relative to cells moving to other zones. This is achieved through what Gordon et al. (2009) call the master *regulator*, the homeodomain protein WUSCHEL (WUS)—homeotic genes encode transcription factors and so control expression of other genes.

Recall from Chapter 4 that regulators work to control the balance between two forces. This master regulator determines how balance between size of the organizing centre (OC) and CZ is maintained for a particular rate of growth activity. Additional processes that, in effect, alter the rate around which the regulator works determine how various conditions can affect growth rate and the progress of cells to form lateral organs. The operation of WUS as a regulator is to balance activity that increases cytokinin production, and so cell division, with a process that reduces WUS itself and so decreases cell division.

Cytokinin produces transcription of the homeodomain protein WUS in the OC (Fig. 5.6)—it has two important effects. It represses the transcription of genes that negatively affect cytokinin signalling (Leibfried et al. 2005)—so tending to increase WUS production in the OC. The WUS protein is transported through plasmodesmata to the cytoplasm of stem cells of the CZ where it also promotes expression of the *CLV3* gene that produces CLV3, a peptide ligand (a binding molecule) that is water soluble and is secreted from stem cells and diffuses

to the OC. The CLV3 peptide binds with the extracellular domain of the CLV1/CLV2 heterodimer and induces autophosphorylation of the cytoplasmic domain of CLV1. Phosphorylated CLV1 binds to downstream effector molecules that repress WUS gene expression (Sharma et al. 2003).

Consequently if there is an increase in the number of stem cells in the CZ relative to cells in the OC then the amount of CLV3 will increase, which in turn will suppress WUS and so decrease stem cell production. This in turn will reduce CLV3 and so in turn will increase expression of WUS (Fig. 5.6).

Gordon et al. (2009) suggest there is a threshold for activation of WUS by cytokinin and that this gives what they term a 'robust' output even if cytokinin signalling fluctuates. This has advantages for the functioning of the regulator. There will certainly be time lags between the perception of signals and metabolic response. If WUS activation was to follow every (small) change then there is the possibility of continuous variation and not reaching stable actuation but possibly oscillating between too much and too little.

The WUS/CLV3 regulator functions within a network of gene controls that modulate effects of the environment on its balance and connect its function as a regulator between stem cells and OC cells with cell growth and development of other zones (Fig. 5.7). Modulation of cytokinin levels in the OC is the principal method of influence. The repression of negative regulators of cytokinin signalling by WUS is through inhibiting ARABIBOPSIS RESPONSE REGULATOR (ARR7), reaction ① in Fig. 5.7, which will have a positive effect on cytokinin production ②. Gaillochet and Lohmann (2015) describe cytokinin levels in SAM as modulated by biosynthetic enzymes produced by the genes *LONELY GUY (LOG)* ③ and *SHOOT MERISTEMLESS (STM)* ④ that activate ISOPENTYL TRANSFERASE (IPT) ⑤ and catabolic enzymes that degrade it (CKX) of the *CYTOKININ OXIDASE/DEHYDROGENASE GENE FAMILY* ⑥.

Cells do not progress from the CZ along a deterministic trajectory but their relative speed is controlled by the HECATE gene (*HEC*) transcription factors ⑦ that interact with both auxin and cytokinin in controlling this progress and also control meristem size and how some environmental

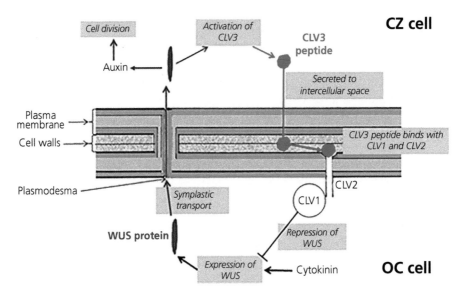

Fig. 5.6 Schematic representation of operation of the WUS/CLV3 master regulator governing the balance of cell numbers in the CZ and OC of SAM. Cytokinin exists within the OC tissue and promotes cell division and the WUS protein is transported actively through plasmodesmata to cells in the CZ where WUS promotes auxin production and so cell division but also activates CLV3 that produces the CLV3 peptide which is secreted into intercellular space. This peptide binds with the extracellular part of the CLV1/CLV2 dimer in OC cells which induces repression of WUS.

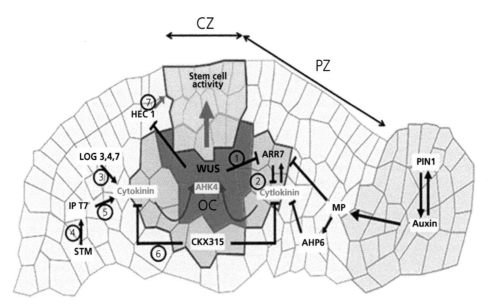

Fig. 5.7 Schematic representation of major interactions in the network of control of SAM growth. OC, red; CZ stem cells, yellow; green and red, cells with cytokinin signalling; blue, developing organ primordium. Numbers within circles label reactions discussed in the text. ⟶ represents a stimulating interaction; ⊣ represents inhibition. Adapted from Gaillochet and Lohmann (2015) with permission from The Company of Biologists under a CC-BY-3.0 license.

factors may affect meristem size (Gaillochet et al. 2017). The rate of lateral organ initiation correlates with SAM size—larger SAMs producing more organs per unit time. Gaillochet et al. (2017) found that transferring plants for 14 days to low light (from 200 to 15 μmol m^{-2}) produced smaller SAM but not smaller cell size and so was affecting cell division. Mutants of *HEC* did not change in meristem size. Plants under low light had markedly reduced cytokinin responses.

Gaillochet et al. suggest that WUS controls the spatial distribution of *HEC* mRNA by repressing its expression and this maintains SAM. The lateral extension of SAM is controlled in part by the balance between cytokinin and auxin. Auxin is synthesized in newly formed tissue of the PZ whereas cytokinin is not. Auxin signals through the AUXIN RESPONSE FACTOR called MONOPTEROS (MP), ⑧ in Fig. 5.7. This has two effects: within the OC it represses ARR7 which reduces CLV3 signalling and thereby tends to increase cytokinin and so WUS; in cells outside of the CZ MP promotes expression of ARABIDOPSIS HISTIDINE PHOSPHOTASE TRANSFER PROTEIN (AHP6) ⑨ which interferes with the cytokinin signal transduction pathway ⑩ and so restricts cytokinin signalling to the centre of SAM. This difference between effects in PZ and CZ is considered by Zhao et al. (2010) as related to an observed gradient in MP RNA.

5.4.2 Initiation of lateral organs

Formation of lateral organs (leaves or flowers) requires initiation of a primordium on the periphery of the apical dome and separated from other primordia. This development is controlled through concentration of auxin at specific points, which results in cell extension, and less extension in regions from which auxin has been conducted.

Auxin, indole acetic acid (IAA) (Taiz et al. 2022, section 19.1), stimulates cell extension through increase in hydrogen ions from H$^+$-ATPase which has the effect of loosening the existing primary cell wall, so enabling an extension response to cell turgor pressure and enabling cellulose to be added into the wall (see the acid growth hypothesis in Taiz et al. 2022, section 2.3). Auxin is produced by cells in the peripheral zone and primordia through auxin being concentrated in a convergence pattern to a position on the shoulder of the SAM.

Auxin is concentrated by its movement between cells in the tissue. The pH outside of the plasma membrane is maintained in an acid condition and auxin is taken into cells by diffusion in the protonated form (IAAH) or by the auxin influx carrier protein AUX1, a proton symporter. Within the cytosol IAA dissociates to the anionic form. The plasma membrane is impermeable to the anion and it accumulates or exits cells by an auxin efflux carrier PIN protein. Crucially PIN proteins are asymmetrically distributed so that auxin movement through a tissue can be directional and operate against a concentration gradient.

Auxin causes a limited reduction in wall stiffness of outer layer cells of the meristem. PIN1 polarities orient in a convergence pattern towards organ initiation sites, indicating a direct role for PIN1 polarity in concentrating auxin locally (Sassi et al. 2014). Convergence sites are located in the outer cell layers. Cells, both in the epidermis and immediately beneath, actively conduct auxin towards the site of the primordium. Conducting auxin away from surrounding cells results in an auxin depletion zone where there is no cell expansion and so no possibility of another primordium forming until new cells are added to the upper part of the peripheral zone that are sufficiently distant that auxin is not effectively conducted away from them so that a new concentration site may be initiated. Localization of PIN1 in convergence patterns is necessary to produce regular organ spacing, but, following initiation of the organ primordium, then polarization towards subepidermal cells is required for development of conducting tissues.

Cellulose fibrils for cell wall growth are produced by the enzyme cellulose synthase that occurs in the plasma membrane (Taiz et al. 2022, section 2.1). Deposition of fibrils into the cell wall by this enzyme is generally oriented perpendicular to the axis of cellular expansion (Hamant et al. 2008). Cellulose synthase is organized in the cell membrane by a functional association with cortical microtubules (Paredez et al. 2006) and newly deposited microfibrils are usually aligned with cortical microtubules that, in the extending cell, are parallel to mechanical stresses in the cell wall (Hamant et al. 2008). Hamant et al. found through live imaging that cortical microtubules at the meristem summit constantly changed their orientation at 1- to 2-h intervals while those in the PZ were dominantly circumferential in orientation and this reflects the types of cell growth taking place.

High auxin levels in a cell produce cell wall loosening and then through pectin linkages if one cell wall is loosened tensile strength in the adjacent neighbouring cell wall will increase. Heisler et al. (2010) suggest that cells target PIN1 protein towards more highly stressed walls so that a neighbouring cell will polarize towards the cell that loosened its wall in response to auxin. Once an organ primordium starts to grow, microtubules become aligned along the boundary of the primordium and Hamant et al. calculated that the directions of principal stress were parallel to the observed microtubule orientations in these different regions of the apical tissue. Experimental application of mechanical stress reorients microtubules towards the axis of the applied force.

5.4.3 Foliage growth

Understanding the control of leaf growth requires reconciling how apparently different requirements for the growth of an organ can be met and how physiological processes with different or even contrasting modes of action may interact. Variation in the rate of leaf production is largely determined by changes in cell division rate in the apical meristem (Poethig 2003). Although we might anticipate differences in size to be the result of differences in cell numbers and/or cell sizes, there seems likely to be tissue level control to ensure this consistent shape.

The rate of leaf production varies with developmental progress of the plant. Méndez-Vigo et al. (2010) identified two distinct phases of *Arabidopsis* grown under short day periods. The first half of vegetative development (respectively ~30 days and ~35 days for two accessions, the laboratory Landsberg *erecta*, *Ler*, and a wild-type Portugal strain, Fei-0) showed rates of leaf production (RLP) increasing rapidly to a near steady-state value. Subsequently RLP increased more slowly. The wild-type strain had higher RLP than Landsberg *erecta* throughout.

From a QTL analysis Méndez-Vigo et al. (2010) conclude that variation in RLP is determined by 10 QTL whose additive effects are under temporal control, some of which are associated with variation in flowering time. They suggest five loci can be interpreted as affecting vegetative or reproductive phase changes, whilst three loci participate more directly in control of RLP.

Within plants of a single species there tends to be consistent leaf shape, although differences in size. Walter and Schurr (2005) note that the primary pattern of leaf growth over time is a diurnal cycle but that this cycle varies between species. Both tobacco (*Nicotiana tabacum*) and *Arabidopsis* (Wiese et al. 2007) have largest growth rates at dawn and minima 12 h later at the day–night transition, while in grasses maximum growth rate may be in the middle of the light period.

Both exogenous and endogenous factors affect leaf growth. Exogenous factors may affect the processes of cell division and cell expansion to different degrees through the supply of resources, such as sugars, but also water which is essential for cell expansion. Endogenous factors, particularly those generated as components of genetically controlled development, may operate through small molecule plant hormones. Based on microarray analysis of young *Arabidopsis* Nemhauser et al. (2006) indicate that seven major hormones regulate different members of gene families, with each hormone acting independently so that endogenous control processes may have an intricate structure.

A number of growth-promoting and growth-restricting factors have been identified in leaf growth, but Krizek (2009) notes that the identified genes 'largely define independent genetic pathways making it difficult to develop an integrated model of organ size control'. (See also Wang et al. 2021.) Pathways that affect organ size involve plant hormones and genes producing proteins that promote growth or genes that produce proteins that can restrict growth (Fig. 5.8). Protein action occurs at different stages of gene → protein regulation and in different ways, as transcription factors, hormone response factors, and interaction with the ubiquitin and protease process. Control can also be exerted through production of miRNA and siRNA. The multiple influences that some components have

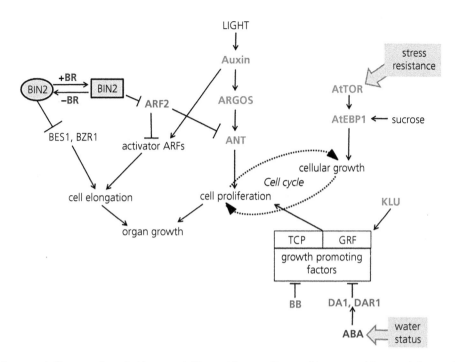

Fig. 5.8 Pathways controlling organ size—plant hormones in blue, proteins promoting growth in green, and those restricting growth in red. Description of major functions is given in the text. Adapted from Krizek (2009) with permission from Elsevier.

through controlling more than one growth process indicate that, overall, the control process should be considered as a network rather than a linear process. The occurrence of what is termed redundancy between genes (Pérez-Pérez et al. 2009), which means that more than one gene may affect a process, also suggests the network approach is appropriate. The network in Fig. 5.8 represents leaf growth as the result of cell proliferation and cell elongation.

Four important features of the control system for organ size (Fig. 5.8) deserve comment:

1. A network of component processes are involved and control cannot be characterized as a straightforward feedback or feedforward system.
2. The network comprises components with different types of action—those tending to increase growth contrasting with those tending to decrease it, the basic structure of a regulator system, particularly those affecting controls of the cell cycle. From this perspective the

network acts to produce an organ of finite size not an organ that will grow as large as available resources might enable. This is an important regulatory requirement in the production of whole plant morphology with repeatable distinctive characteristics.
3. As a number of authors have commented, the controls on organ size have different genes. Furthermore, controls operate through a range of processes involved in protein formation and its regulation. This variation may reflect both the nature of the gene being controlled and where relevant the characteristics of the influencing process and how it is signalled.
4. A number of processes are termed 'redundant' in the literature. This should be taken to mean that two (or more) processes act on the same process but not simply that one or the other could be dispensed with (Kafri et al. 2009). Redundancy in engineering typically involves the replication of components to protect against the consequences of failure.

Cell proliferation (Fig. 5.8) is the result of the cell cycle involving both cell growth and cell division which is under the control of cell cycle genes that are modulated by transcription factors. Gonzalez et al. (2012) note that seven of the nine members of the GROWTH-REGULATING FACTOR (GRF) family are targeted by *miR396*. Mutational analysis has shown that the TEOSINTE BRANCHED/CYCLOIDEA/PCF (TCP) transcription factor family can affect, slightly to strongly, shape from flat to crinkly. These genes are regulated by *miR319*. Overexpression of *miR319* prolongs mitotic activity along the leaf margin.

One way that auxin may control leaf size is through the transcription factor auxin-regulated gene (*ARGOS*) which acts upstream of *AINTEGUMENTA (ANT)*. Mutations in either *ARGOS* or *ANT* result in smaller size. Krizek (2009) notes that *ANT* expression may be negatively regulated by AUXIN RESPONSE FACTOR 2 (ARF2) which is a repressor of organ growth, possibly through action as a transcriptional repressor.

BIN2 is a brassinoid (BR) regulated kinase that can phosphorylate ARF2, stopping its ability to repress transcription of *ANT* (Vert et al. 2008). ARF2 may compete with activator ARFs so its inactivation may allow positive ARF response to affect cell elongation. Vert et al. describe the control: 'in effect, BRs would release a brake from such coregulated genes, and auxin would provide gas to drive increased expression through release of repression on activator ARFs'. Krizek (2009) suggests that BRs may affect the target specificity of BIN2, switching preference from (BRI1-EMS SUPPRESSOR1 (BES1) and BRASSINAZOLE RESISTANT1 (BZR1), which action it attenuates, to ARF2. In the presence of both hormones there would be increased and positively acting ARFs and BES1/BZR1.

The *TARGET OF RAPAMYCIN (TOR)* pathway provides important connection between the environment and leaf growth (Krizek 2009). The *TOR* pathway is a conserved signalling network in mammals, insects, and yeast. The *Arabidopsis* ortholog *Arabidopsis* TOR gene (*AtTOR*) works in conjunction with Ell Binding Protein (AtEBP1), a multifunctional protein involved in translational control and transcriptional regulation (reviewed in Krizek 2009). *AtTOR* expression affects the ability of transgenic plants to withstand osmotic stress and appears to regulate both cellular biomass and stress resistance. *Arabidopsis* plants with elevated or reduced expression of EBP1 have, respectively, larger or smaller leaves. Its expression is upregulated by sugar and, at the level of protein stability, by auxin.

A primary spatial pattern in leaf growth is a decline in growth rate from base to tip. Although several metabolites involved with growth have pronounced diurnal variations in young but not in full grown leaves the most consistent correlation between spatial distribution of growth and metabolite concentrations was found by Walter and Schurr (2005) for the ratio between glucose and fructose concentration which declines from leaf base to tip in a similar manner to the fall in growth rate.

However, Anastasiou et al. (2007) define a gene and its mode of action that controls organ size. With the convention of naming genes after their loss-of-function phenotype, this gene is termed *KLU*, the inverse of the comic monster Hulk, and is found to operate in leaves, sepals, and petals. *KLU* stimulates organ growth by preventing cell proliferation from stopping. Cell proliferation ceases earlier in *klu* mutants, both in time and accumulated cell number, suggesting a cell-number-dependent process. *KLU* in wild type counteracts this, leading to a less rapid decline of proliferation at later stages. Crucially they found that *KLU* is expressed outside of the region of cell proliferation and *KLU* expression continued when cells had ceased to proliferate, indicating that it is not simply downregulation of *KLU* that causes cells to stop proliferating.

Anastasiou et al. found that *KLU* acts independently of other genes listed in Fig. 5.8 and its growth-stimulating signal is distinct from characterized phytohormones. They suggest *KLU* generates a growth-stimulating compound that is mobile in the tissue and that members of the cytochrome family are involved. In both petals and leaves *KLU* is expressed in limited regions, around the periphery or base respectively, and increases more slowly than organ size as a whole, so that the KLU-dependent signal will be diluted as the organ grows and as soon as it falls below a certain concentration it will no longer sustain cell proliferation. Anastasiou et al. suggest: 'a signal source that grows more slowly than the primordium as a whole can only sustain proliferation up to a certain point, ensuring that growth terminates once the organ has reached its target size'.

Krizek (2009) describes genes that repress growth through encoding proteins associated with the ubiquitin-proteasome degradation system with mutants that produce larger organs. BIG BROTHER (BB) (Fig. 5.8) restricts the period of cell proliferation possibly through degradation of stimulating factors. The *DA1* (DA means 'large' in Chinese) is a member of a small family of DA1-related (DAR) proteins that also restrict the period of growth, but *DA1* and *BB* double mutants have an additive effect. However, while *BB* expression is independent of plant hormones, *DA1* expression is upregulated after plant exposure to abscisic acid (ABA) (Li et al. 2008). ABA is a monitor of water status (Christmann et al. 2006).

5.5 Discussion

Useful analogies can be made between control of processes involved in plant growth and control systems in engineering. The circadian clock produces feedforward control; the WUS/CLV3 system acts as a regulator; the PIN1/auxin system acts as a positive feedback for auxin accumulation which produces cell wall loosening and cellulose microfibre increment. In each of these examples controls are embedded within a network of nuclear control that acts as an interface with the environment and, as in the case of the apical meristem, to integrate the functions. Of course in engineering applications, as tasks have increased in complexity, additional controls have led to construction of networks; the cruise control regulator for vehicles may be superseded by autonomous vehicle technology that involves multiple sensing, signalling, and drive systems.

As a plant grows its constituent modules and the apical meristems that produce them become spatially distributed with a substantially regular pattern. Modules and meristems throughout the plant can experience different environments and come to have different status and relationships with other modules on which they depend (discussed further in Chapter 7). However, the circadian clock ensures nuclear control at the local (cellular) level and an integrated response to the local environment in that physiological systems are gated to respond to conditions in the diurnal cycle that are most suitable for them. The physiology of plant apices maintains pluripotency. Contrasting conditions between the cytosol and cell wall and intracellular space, combined with active transport of signal molecules in some conditions, are integral parts of the tissue control systems maintaining the apical meristem and initiating foliage.

Consistency in development and growth is produced by the repeating relationship between components within the constraints and opportunities of the environment. It has some characteristics of distributed control systems used in engineering but with the important difference that there is no central control, at least not as seen in most animals that have central nervous and circulatory systems. Active centres of development (meristems) are spatially separated and have some autonomous activity, particularly in response to environmental variation, but they are also dependent upon connections to other organs for some resources and components of their development. Lachowiec et al. (2016) define the concept of *robustness* as: 'The ability of an organism to maintain a specific phenotypic value or state in the face of environmental and genetic permutations.' They suggest it arises from: 'the structure of genetic networks, the specific molecular functions of the underlying genes and their interactions'. In particular they suggest that apparent *redundancy* in genes governs robustness. This is discussed further in Chapter 7.

5.6 References

Anastasiou, E., Kenz, S., Gerstung, M., et al. 2007. Control of plant organ size by KLUH/CYP78A5-dependent intercellular signaling. *Developmental Cell*, 13, 843–856.

Bell, A. D. 2008. *Plant Form*. Portland, Oregon, Timber Press.

Christmann, A., Moes, D., Himmelbach, A., Yang, Y., Tang, Y. and Grill, E. 2006. Integration of abscisic acid signalling into plant responses. *Plant Biology*, 8, 314–325.

Collings, D. A., Carter, C. N., Rink, J. C., Scott, A. C., Wyatt, S. E. and Allen, N. S. 2000. Plant nuclei can contain extensive grooves and invagination. *The Plant Cell*, 12, 2425–2439.

Endo, M., Shimizu, H., Nohales, M. A., Araki, T. and Kay, S. A. 2014. Tissue-specific clocks in *Arabidopsis* show asymmetric coupling. *Nature*, 515, 419–422.

Farré, E. M. and Weise, S. E. 2012. The interactions between the circadian clock and primary metabolism. *Current Opinion in Plant Biology*, 15, 293–300.

Filichkin, S. A., Ghislain Breton, H. D., Priest, P. D., et al. 2011. Global profiling of rice and poplar transcriptomes highlights key conserved circadian-controlled pathways and cis-regulatory modules. *PLoS ONE*, 6, 6:e16907.

Gaillochet, C. and Lohmann, J. U. 2015. The never-ending story: from pluripotency to plant developmental plasticity. *Development*, 142, 2237–2249.

Gaillochet, C., Stiehl, T., Wenzl, C., et al. 2017. Control of plant cell fate transitions by transcriptional and hormonal signals. *eLife*, 6, e30135.

Gonzalez, N., Vanhaeren, H. and Inze, D. 2012. Leaf size control: complex coordination of cell division and expansion. *Trends in Plant Science*, 17, 332–340.

Gordon, S. P., Chickarmane, V. S., Ohno, C. and Meyerowitz, E. M. 2009. Multiple feedback loops through cytokinin signaling control stem cell number within the *Arabidopsis* shoot meristem. *Proceedings of the National Academy of Sciences of the United States of America*, 106, 16529–16534.

Graf, A. and Smith, A. M. 2011. Starch and the clock: the dark side of plant productivity. *Trends in Plant Science*, 16, 169–175.

Greenham, K. and McClung, C. R. 2015. Integrating circadian dynamics with physiological processes in plants. *Nature Reviews Genetics*, 16, 598–610.

Gyllenstrand, N., Karlgren, A., Clapham, D., et al. 2014. No time for spruce: rapid dampening of circadian rhythms in *Picea abies* (L. Karst). *Plant and Cell Physiology*, 55, 535–550.

Hamant, O., Heisler, M. G., Jonsson, H., et al. 2008. Developmental patterning by mechanical signals in Arabidopsis. *Science*, 322, 1650–1655.

Harmer, S. L., Hogenesch, J. B., Straume, M., et al. 2000. Orchestrated transcription of key pathways in *Arabidopsis* by the circadian clock. *Science*, 290, 2110–2113.

Haydon, M. J., Mielczarek, O., Robertson, F. C., Hubbard, K. E. and Webb, A. A. R. 2013. Photosynthetic entrainment of the *Arabidopsis thaliana* circadian clock. *Nature*, 502, 689–692.

Heisler, M. G., Hamant, O., Krupinski, P., et al. 2010. Alignment between PIN1 polarity and microtubule orientation in the shoot apical meristem reveals a tight coupling between morphogenesis and auxin transport. *PLoS Biology*, 8, e1000516.

Inoue, K., Araki, T. and Endo, M. 2017. Integration of input signals into the gene network in the plant circadian clock. *Plant and Cell Physiology*, 58, 977–982.

James, A. B., Monrea, A., Nimmo, G. A., et al. 2008. The circadian clock in Arabidopsis roots is a simplified slave version of the clock in shoots. Science, 322(5909), 1832–1835.

Kafri, R., Springer, M. and Pilpel, Y. 2009. Genetic redundancy: new tricks for old genes. *Cell*, 136, 389–392.

Khan, S., Rowe, S. C. and Harmon, F. G. 2010. Coordination of the maize transcriptome by a conserved circadian clock. *BMC Plant Biology*, 10, 126.

Krizek, B. 2009. Making bigger plants: key regulators of final organ size. *Current Opinion in Plant Biology*, 12, 17–22.

Lachowiec, J., Queitsch, C. and Kliebenstein, D. J. 2016. Molecular mechanisms governing differential robustness of development and environmental responses in plants. *Annals of Botany*, 117, 795–809.

Leibfried, A., To, J. P. C., Busch, W., et al. 2005. WUSCHEL controls meristem function by direct regulation of cytokinin-inducible response regulators. *Nature*, 438, 1172–1175.

Li, Y. H., Zheng, L. Y., Corke, F., Smith, C. and Bevan, M. W. 2008. Control of final seed and organ size by the DA1 gene family in *Arabidopsis thaliana*. *Genes and Development*, 22, 1331–1336.

Louveaux, M., Rochette, S., Beauzamy, L., Boudaoud, A. and Hamant, O. 2016. The impact of mechanical compression on cortical microtubules in *Arabidopsis*: a quantitative pipeline. *Plant Journal*, 88, 328–342.

Méndez-Vigo, B., De Andrés, M. T., Ramiro, M., Martínez-Zapater, J. M. and Alonso-Blanco, C. 2010. Temporal analysis of natural variation for the rate of leaf production and its relationship with flowering initiation in *Arabidopsis thaliana*. *Journal of Experimental Botany*, 61, 1611–1623.

Nemhauser, J. L., Hong, F. X. and Chory, J. 2006. Different plant hormones regulate similar processes through largely nonoverlapping transcriptional responses. *Cell*, 126, 467–475.

Novák, B. and Tyson, J. J. 2008. Design principles of biochemical oscillators. *Nature Reviews. Molecular Cell Biology*, 9, 981–991.

Para, A., Farré, E. M., Imaizumi, T., Pruneda-Paz, J. L., Harmon, F. G. and Kay, S. A. 2007. PRR3 is a vascular regulator of TOC1 stability in the *Arabidopsis* circadian clock. *The Plant Cell*, 19, 3462–3473.

Paredez, A. R., Somerville, C. R. and Ehrhardt, D. W. 2006. Visualization of cellulose synthase demonstrates functional association with microtubules. *Science*, 312, 1491–1495.

Pérez-Pérez, J. M., Candela, H. and Micol, J. L. 2009. Understanding synergy in genetic interactions. *Trends in Genetics*, 25, 368–376.

Poethig, R. S. 2003. Phase change and the regulation of developmental timing in plants. *Science*, 301, 334–336.

Sassi, M., Ali, O., Boudon, F., et al. 2014. An auxin-mediated shift toward growth isotropy promotes organ formation at the shoot meristem in *Arabidopsis*. *Current Biology*, 24, 2335–2342.

Seki, M., Ohara, T., Hearn, T. J., et al. 2017. Adjustment of the *Arabidopsis* circadian oscillator by sugar signalling dictates the regulation of starch metabolism. *Scientific Reports*, 7, 8305.

Sharma, V. K., Ramirez, J. and Fletcher, J. C. 2003. The *Arabidopsis* CLV3-like (CLE) genes are expressed in diverse tissues and encode small secreted proteins. *Plant Molecular Biology*, 51, 415–425.

Shim, J. S. and Imaizumi, T. 2015. Circadian clock and photoperiodic response in *Arabidopsis*: from seasonal flowering to redox homeostasis. *Biochemistry*, 54, 157–170.

Taiz, L., Zeiger, E., Møller, I. M. and Murphy, A. 2022. *Plant Physiology and Development*, Seventh Edition. Oxford, Oxford University Press.

Takahashi, N., Hirata, Y., Aihara, K. and Mas, P. 2015. A hierarchical multi-oscillator network orchestrates the *Arabidopsis* circadian system. *Cell*, 163, 148–159.

Truskina, J. and Vernoux, T. 2018. The growth of a stable stationary structure: coordinating cell behavior and patterning at the shoot apical meristem. *Current Opinion in Plant Biology*, 41, 83–88.

Vert, G., Walcher, C. L., Chory, J., and Nemhauser, J. L. 2008. Integration of auxin and brassinosteroid pathways by auxin response factor 2. *Proceedings of the National Academy of Sciences of the United States of America*, 105, 9829–9834.

Walter, A. and Schurr, U. 2005. Dynamics of leaf and root growth: endogenous control versus environmental impact. *Annals of Botany*, 95, 891–900.

Wang, H., Kong, F. and Zhou, C. 2021. From genes to networks: The genetic control of leaf development. *Journal of Integrative Plant Biology*, 63, 1181–1196.

Webb, A. A. R. and Satake, A. 2015. Understanding circadian regulation of carbohydrate metabolism in *Arabidopsis* using mathematical models. *Plant and Cell Physiology*, 56, 586–593.

Wiese, A., Christ, M. M., Virnich, O., Schurr, U. and Walter, A. 2007. Spatio-temporal leaf growth patterns of *Arabidopsis thaliana* and evidence for sugar control of the diel leaf growth cycle. *New Phytologist*, 174, 752–761.

Zhao, Z., Andersen, S. U., Ljung, K., et al. 2010. Hormonal control of the shoot stem-cell niche. *Nature*, 465, 1089–1092.

CHAPTER 6

Stability of the photosynthesis system

6.1 Introduction

In the example of seasonal and diurnal variation in photosynthesis (Chapter 3) values of A_{max} were generally *less* when light flux was generally *greater*. Processes associated with protection and repair of the photosynthesis system were identified that may contribute to this (Fig. 3.9). This chapter describes components of their control systems and illustrates their importance for stability of the photosynthesis system.

Importance of the decline in photosynthesis at high light flux has been emphasized. Zavafer et al. (2015) suggest 'photosynthesis is intrinsically a suicidal process since exposure to light will also cause inhibition of PSII machinery'. Photochemical generation of the electron stream can also generate reactive oxygen species (ROS) that cause damage to proteins of the photosynthesis system. Oelze et al. (2008) suggest: 'Regulation of the photosynthetic apparatus between efficient energy conversion at low light and avoidance of overreduction and damage development at excess light resembles dangerous navigating between Scylla and Charybis.'

Control of photosynthesis should be considered with regards to maintenance of a chemical environment that ensures stability of the structure. This involves processes that can protect against, or reduce, potential damage as well as repair damage that occurs. It also involves maintenance of physiological systems in a state that CO_2 fixation can increase following periods of low light rather than being stalled. Both structure and biochemical properties of the system are involved.

6.2 Variation in chloroplast and thylakoid membrane structure

Component processes of photosynthesis are typically represented to illustrate progression of light reactions and their relationship with the Calvin–Bensson–Bassam (CBB) cycle (Fig. 6.1). However, the functioning of chloroplasts, both in carbon fixation and photoprotection and repair from damage, is intimately related to their structure, which changes as light flux changes in ways that facilitate the repair processes of PSII and functioning of cyclical electron flow (CEF) through PSI.

Thylakoid membranes enclose a single, continuous lumen and are folded, forming two distinct domains: grana, which comprise 80% of the membrane (Albertsson 2001) and appear as stacks of circular structures, and stroma lamella, which are unstacked lamella and connect the grana stacks (Dekker and Boekema 2005). Typically PSII exists as dimers embedded in the grana thylakoid membrane where it is associated with light harvesting complexes of chlorophylls (LHCII) (Fig. 6.2). PSI and its associated light harvesting complexes (LHCI) are located in the stroma thylakoid membrane along with ATPase. Cytochrome b_6/f complexes are distributed between the grana and stroma thylakoids.

Grana discs have diameters of ~300–600 nm and several dozens of membranes may be stacked together (Dekker and Boekema 2005); distance between membranes on the stromal side is ~2 nm (distance A in Fig. 6.2) but the distance across the thylakoid lumen (distance B in Fig. 6.2) is greater and so accommodates protrusions from the PSII core and cytochrome b_6f.

The Dynamics of Plant Growth. E. David Ford, Oxford University Press. © E. David Ford (2023). DOI: 10.1093/oso/9780192867179.003.0006

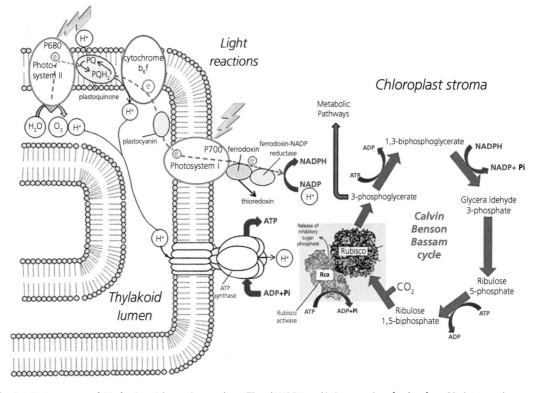

Fig. 6.1 Basic processes of CO_2 fixation. Light reactions produce ATP and NADPH used in incorporation of carbon from CO_2 into organic compounds by RuBisCo; associated enzymes regenerate the acceptor molecule RuBP. Thylakoid membranes enclose the thylakoid lumen and comprise lipid molecules (yellow circles with tails) within which are embedded chlorophylls and proteins of PSII and PSI. Embedded macromolecules are only shown in one thylakoid membrane but occur throughout the thylakoids and in particular arrangements discussed in the text. Excited electrons (red dashed line) follow what is termed the Z scheme of electrical potential—after initial excitation they lose electrical potential as they are carried from PSII to PSI where they are further excited sufficiently to produce ferredoxin and NADPH released into the chloroplast stroma. As a result, hydrogen ions accumulate in the thylakoid lumen, producing a pH difference across the membrane which drives the production of ATP by ATP synthase. Cyclical electron transfer can occur through PSI when electrons are donated back to plastoquinone; this generates H^+ accumulation in the thylakoid lumen and so ATP formation but not ferredoxin or NADPH. Plastoquinone moves through the thylakoid membrane and plastocyanin through the thylakoid lumen. ATP is utilized in activation of RuBisCo by rca and in CBB reactions. RuBisCo activation is a continuous process to remove sugar phosphate molecules that can block its reaction sites. NADPH is used in reactions of the CBB and, along with ATP, other metabolic reactions in the chloroplast. Both light reactions and the CBB are described by Taiz et al. (2022, in their chapters 9 and 10 respectively).

Stacking provides PSII with a large functional antenna—excitation energy can flow within and between two stacked membranes until being quenched in an open PSII reaction centre. The vertical repeat distance within grana (Fig. 6.2) varies between 14 and 24 nm according to environment and species (Dekker and Boekema 2005).

CEF operates efficiently at the onset of illumination, which can only occur if the linear and cyclic electron transfer chains are physically separated from each other (Dekker and Boekema 2005). PSI are located in the stroma thylakoid membranes and CEF operates removed from the grana, with b_6f also located in the stroma lamella. Dekker and Boekema suggest that ferredoxin-NADP reductase (FNR) plays a key role in discriminating between linear and cyclic routes by binding to PSI for linear electron transport and to b_6f for cyclic electron transport. ATPase is located only in the stroma lamellae and at the ends of stacked grana (Fig. 6.2). The rotating head of that enzyme protrudes from the thylakoid membrane and could not be accommodated within grana stacks.

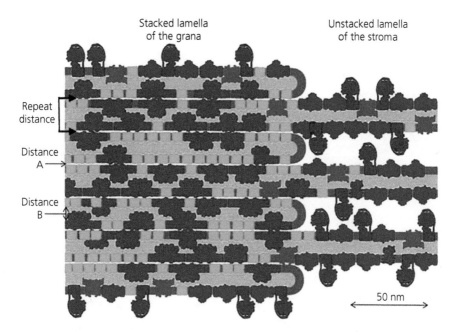

Fig. 6.2 Diagrammatic representation of the green plant thylakoid membrane. Dimeric PSII–LHCII supercomplexes and PSII monomers in dark green; single LHCII trimers bright green; PSI dark blue; cytochrome b6/f dimers orange; ATP synthase red. The acidic lumen is bright blue. The diagram represents approximately half an average grana stack in the horizontal dimension. Under some conditions, the number of cytochrome b6/f dimers in the grana can be higher than shown. Reproduced from Dekker J. P. and Boekema E. J. (2005) with permission from Elsevier.

The size and structure of thylakoids vary with conditions, particularly light flux. This may affect rate of transfer of electron carriers and macromolecules (reviewed by Kirchhoff 2013). Under low light the number of membranes in grana increases, providing greater potential activation of chlorophyll. Two structural changes can occur under high light stress: vertical unstacking and lateral shrinkage of grana, both of which may reduce transfer of activation between pigment molecules. The width of the thylakoid lumen expands in light (Fig. 6.3) which Kirchhoff et al. (2011) suggest is the result of the proton motive force that develops with light and induces osmotic swelling by an increase in Cl⁻ concentration.

Both the membrane and lumen of thylakoids are densely packed with proteins—the lumen contains around 80 different proteins involved in the regulation, protein degradation, and maturation processes. Kirchhoff (2014) suggests that dense packing, particularly in grana membranes, facilitates energy capture by chlorophyll and associated pigments but restricts diffusion to local regions. Swelling of the lumen will facilitate transfer of plastocyanin and xanthophylls.

6.3 Control systems

Three nodes are focal points in control of the photosynthesis system (Fig. 6.4):

- Node 1: control of the flow of activated electrons that initiate LEF.
- Node 2: control of the NADPH/ATP ratio, CEF, and redox reactions.
- Node 3: controls of the CBB cycle that sustain its function and retards changes in chloroplast chemistry.

Nodes 1 and 2 both act as governors. However, they do not have fixed set points of control but function to produce an activated electron flow, reduce possibility of an excess flow, and adjust the balance of metabolites and ROS from node 2. Functions at node

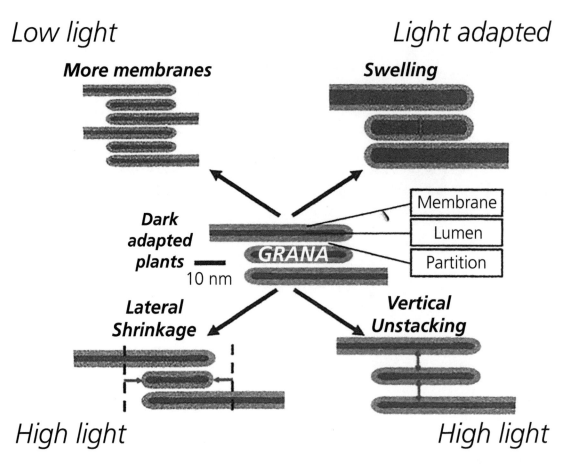

Fig. 6.3 Structural changes in thylakoid grana membranes and their arrangement in chloroplasts. The central diagram represents structure in dark adapted non-stressed plants. For plants growing under low light (upper left) there is increase in numbers of membranes in the grana, while light adapted plants (upper right) have increased lumen space. High light produces two changes: a lateral shrinkage of grana (lower left) and vertical unstacking (lower right). Reproduced from Kirchhoff H. (2013) with permission from Springer Nature.

3 preserve RuBisCo from proteolysis, maintain components of the CBB cycle to readily resume their functions when light increases, and contribute to maintenance of chemical balance and conservation of fixed carbon. The capacity of the CBB to function is *regulated* but its particular outputs are not.

6.3.1 Node 1: governor of activated electrons from PSII

Two processes are involved in control of initiation of linear electron flow (LEF). One is, obviously, to generate LEF. The other is to protect against the potential effects of large increases in light flux; these might be rapid, taking place over a few seconds, and of varied duration, but are sometimes sustained for considerable periods. This control involves the two subsystems of photoprotection against energy reaching PSII and PSII damage and repair, which operate in series.

6.3.1.1 Generation of LEF and potential for damage to PSII

PSII is a large membrane–protein complex extending through the thylakoid membrane into the thylakoid lumen (Fig. 6.5) (Zavafer et al. 2015), as follows: (i) the reduction centre is a pair of chlorophylls, P680, a pheophytin molecule and a

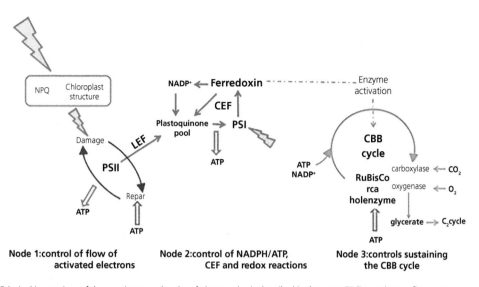

Fig. 6.4 Principal interactions of three major control nodes of photosynthesis described in the text. LEF, linear electron flow; NPQ, non-photochemical quenching.

cytochrome, all embedded in protein subunits, D1 and D2. Photons captured by the light harvesting antenna complex, LHCII, of carotenoids and chlorophylls result in charge separation, i.e. excitation of an electron to a higher energy level, and its transport resulting in oxidized P680$^+$ and reduced pheophytin. (ii) The acceptor has two quinone molecules embedded in the protein subunit that receive electrons from the pheophytin. (iii) The donor side has a metal cluster, Mn_4CaO_5, and a tyrosine (Tyr). The Mn_4CaO_5 donates electrons to reduce P680$^+$ and after accumulating four oxidizing equivalents two water molecules are oxidized to give molecular oxygen, four protons that contribute to acidification of the thylakoid lumen, and four electrons that reduce Mn_4CaO_5 back to its resting state.

Two processes may result in photodamage to PSII, although Zavafer et al. (2015) suggest both may operate as a hybrid system (Oguchi et al. 2011). One is that if PSII absorbs more light than used in photochemistry, or there is some restriction to forward electron transport from the acceptor side, then ROS may be formed. Blocked electron transport can lead to formation of the triplet excited state of P680 via charge recombination with pheophytin to give ^{3}P680 that interacts with molecular oxygen to form a highly reactive singlet oxygen, $^1O_2$ (Pinnola and

Bassi 2018), which damages the D1 protein (Vass 2012).

The other process involves direct absorption of light by the Mn_4CaO_5 cluster leading to formation of comparatively long-lived oxidized Tyr-Z$^+$ and P680$^+$ which, in the absence of sufficient electron flow from the Mn_4CaO_5, cannot be reduced, resulting in high oxidizing power that can damage proteins. UV light can also damage Mn_4CaO_5, leading to imbalance of electron flow. This would explain the consistent experimental result that inactivation of PSII is proportional to radiance, i.e. that it does not depend upon excess light greater than that used in production of NADPH$^+$ and ATP for photosynthesis.

Zavafer et al. (2015) suggest that explanation of the action spectrum of the inactivation of PSII requires that both of these processes operate. Photoinhibition is the net result of damage and repair. Damage to PSII occurs at all light flux and there is a repair system for it, discussed later in section 6.3.1.3.

There may be differences between plant types in the rate of these reactions; Shirao et al. (2013) note that the electron outflow from PSII is more rapid in gymnosperms than angiosperms and which have a higher capacity for O_2-dependent electron flow.

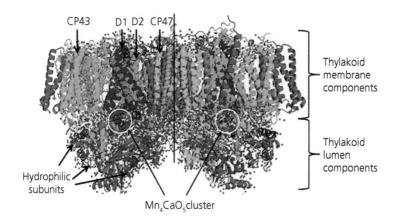

Fig. 6.5 Structure of the PSII dimer viewed from the side. Components that are embedded in the thylakoid membrane are towards the top and hydrophilic components that are in the lumen towards the bottom. The central line indicates the axis of division between two monomers. D1 and D2 form the reaction centre core of PSII, with which electron-transfer and water-splitting reactions are associated. Surrounding the D1 and D2 subunits are the CP47 and CP43 subunits (CP stands for chlorophyll protein) that bind a number of chlorophyll molecules that function in light harvesting. Reproduced from Shen J.-R. H. (2015) "The structure of Photosystem II and the mechanism of water oxidation in photosynthesis". *Annual Review of Plant Biology* 66:23–48, with permission from Annual Reviews.

6.3.1.2 Control of photoprotection

Photoprotection is the process whereby photon passage to the PSII reduction centre is lessened and so less photodamage may occur. A field example is given in Chapter 3.

Two processes, both involving molecular and supramolecular reorganization of LHCII, can be involved: non-photochemical quenching (NPQ) and state transition in chlorophyll between PSII and PSI (section 6.3.2). Both are influenced by products of photochemistry so that photoprotection is controlled in relation to light flux, at least to some extent. NPQ is the major process in photoprotection (Horton et al. 1996) and consists of the loss of excitation energy as heat, i.e. energy quenching (qE) (Latowski et al. 2011) rather than its transfer to the PSII reaction centre.

PSII is surrounded by light harvesting complexes (LHCII). The structure of this association can vary somewhat, but in plants adapted to low light the dominant form is the super complex $C_2S_2M_2$ (Su et al. 2017; van Bezouwen et al. 2017). Ruban (2016) provides a scheme for the structure and organization of the PSII reaction centre and chlorophyll (Fig. 6.6) and describes how that structure changes in the NPQ state. The minor LHCII, listed as CP24,

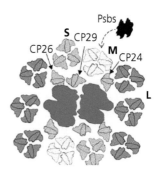

Fig. 6.6 Structure and association of PSII dimmer (red) and its antenna components. Major LHCII trimers are strongly (S, grey), moderately (M, light blue), or loosely (L, dark blue) bound to the core trimers CP26, CP 29, and CP24. Non-pigment protein (PsbS) dimer is shown with a dashed line pointing to an interaction site in the dark suggested by Ruban (2016). Reproduced from Ruban H. V. (2016) with permission from Oxford University Press.

CP26, and CP29 (CP, chlorophyll protein), surround the PSII dimer and form a structural and functional bridge between the core antenna of the PSII dimer and the major LHCII antenna, which are trimers and distinguished according to their binding strengths: strong S, moderate M, loose L.

Two types of molecules are involved synergistically in qE: xanthophyll, which can exist in three forms—violaxanthin, antheraxanthin, and zeaxanthin—and a non-pigment protein (PsbS) with which the xanthophyll is associated (Roach and Krieger-Liszkay 2012). In low light flux violaxanthin is the dominant xanthophyll form, but as light flux increases this is converted enzymatically and progressively to antheraxanthin and then zeaxanthin. The enzyme violaxanthin de-epoxidase converts violaxanthin to antheraxanthin and antheraxanthin to zeaxanthin using ascorbate in both reactions. The enzyme zeaxanthin epoxidase performs the reverse reactions which require NADPH. The S_1 excited energy state of violaxanthin is greater than that of Chl a whereas that of zeaxanthin is less than that of Chl a (Latowski et al. 2011) so that charge transfer quenching can occur from Chl a to zeaxanthin. The photochemical process results in production of protons which affect three components of the photoprotection system: the enzyme violaxanthin de-epoxidase, the PsbS, and the LHCII antenna.

The structural organization between PSII and its LCHII antenna components is crucial to photoprotection which requires disruption of their close contact that would otherwise produce energy transfer. The protein, PsbS, is aggregated with the LHCII antennae complexes (Sacharz et al. 2017) (Fig. 6.6)

and has a synergistic effect on photoprotection (Ruban 2016). Sacharz et al. (2017) suggest PsbS promotes disassociation of LHCII and rearrangement towards the photoprotective state and this process is *independent* of reaction centres. (See also Kereïche et al. 2010.)

Ruban (2016) suggests the non-photochemical 'quencher is simply "born" out of the change in conformation triggered by protonation (Formaggio et al. 2001)'. Fig. 6.7 illustrates aggregations of core dimers and LHCII trimers. PsbS controls dissociation of the portion of the super complex containing LHCII, CP24, and CP29. Clustering of LHCII is increased by zeaxanthin and PsbS. Sacharz et al. (2017) suggest that 'presence of zeaxanthin during NPQ promotes the formation of large LHCII oligomers that contain not only PsbS but CP29 complexes'.

Under conditions of high light the components of the system (the LHCIIs, xanthophylls, PsbS) respond by separation of types and within types by aggregations, resulting in increased energy transfer as heat and reduced energy transfer to reaction centres. This is a fast response system. Ruban (2016) suggests that 'ΔpH provides feedback control over light harvesting efficiency in the photosynthetic membrane'. ΔpH is certainly a component of a

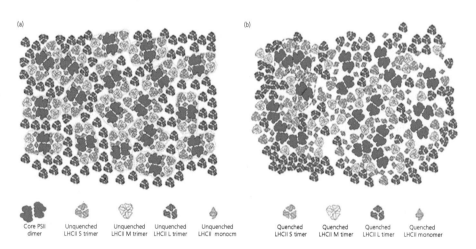

Fig. 6.7 Schematic representations of PSII and LHCII trimers in (a) dark and (b) NPQ conditions, proposed by Ruban (2016) and based on cryo electron tomography (Kouril et al. 2011). $C_2S_2M_2$ complexes are surrounded by yellow lines in (a) and three with red dashed lines under NPQ in (b). Quenched trimers and monomers are outlined with red. Reproduced from Ruban H. V. (2016) with permission from Oxford University Press.

control system, but crucially this is not closed-loop *feedback control*. There is certainly active control involving expenditure of energy to achieve certain conditions, notably through the enzymes of the xanthophyll cycle, but there is no monitoring of the condition that has been reached and so no matching of any current condition against a set point. An appropriate description is that photoprotection involves open-loop control.

Ruban (2017) suggests that the photoprotective component of NPQ in relation to light intensity can be considerably higher than the minimum required for protection, but it is not clear what that minimum might be. However, Ruban notes that photoinhibition is a common occurrence so that NPQ may be limited in its effectiveness. This limitation could be due to the control process which results in spatial reorganization of molecules and the supramolecular structure. Although, as Ruban mentions, under high light flux there is separation of LHCII trimers from PSII dimmers there is a necessarily stochastic component to this reorganization and possibly also in the association between violaxanthin and chlorophyll. There may be two reasons for this.

First, the relative amounts of PSII, LHCII, and PsbS-xanthophyll are likely set in the acclimation process of foliage development and these may not match the ideal proportions for any particular dark $\rightarrow$ light transition. Second, it is reasonable to assume that the spatial evenness of PSII dimers within a matrix of LHCII trimers makes effective use of low light flux, but that very evenness may constrain the extent to which clumping can be achieved under high light conditions.

The important characteristics of the NPQ system are the rapidity of its response and that it is reversible, and together these characteristics provide some response to the inevitable fluctuations in light experienced by foliage.

6.3.1.3 Effects of damage to PSII and control of repair

Degradation and renewal of PSII D1 protein and the functional components bound within it is a continuous process occurring in light fluxes that would not normally be considered as high. For *Brassica napa* in growth chambers at 350 μmol m^{-2} s^{-1} Sundby et al.

(1993) estimated a turnover rate of D1 protein at 50% in 2 h, i.e. 97% of D1 protein is renewed within 10 h. Murata et al. (2012) suggest that increased photoinhibition with increased light flux is due to deceleration of repair, not acceleration of damage, and this is the result of interference by ROS which inhibits protein synthesis.

The PSII repair cycle works in conjunction with structural changes of the thylakoid membranes that play a significant role in the diffusion of the photosynthetic proteins, kinases, phosphatases, and proteases. It involves removal of D1 from PSII, its degradation, and then synthesis and insertion of a replacement protein along with its functional components that are bound to it. This requires a series of enzymatic steps that are spatially separated through movement of PSII within the thylakoid membrane. This movement is enabled by swelling of thylakoids at the grana margins and some unstacking of grana (Puthiyaveetil et al. 2014; Yoshioka-Nishimura 2016).

The first stage is monomerization of the PSII complex and detachment of the PSII core from LHCII in the grana thylakoids (Fig. 6.8). This is achieved through phosphorylation of PSII core subunits catalyzed by STN8 kinase that is linked to the destacking and lateral shrinkage of grana (Järvi et al. 2015). Light-induced swelling and unstacking of thylakoid membranes enables lateral migration of PSII from the grana to the grana margins and phosphorylation of PSII core proteins facilitates accessibility of repair processes.

Once moved from the grana thylakoid the PSII monomer is dephosphorylated which is essential for degradation. This is processed by two categories of proteases, Deg and FtsH, concentrated in the margins of the grana thylakoid. Acidification of the thylakoid lumen is important in activation of Deg1 and also stimulates a required oligomerization of FtsH. Yoshioka-Nishimura (2016) points out that FtsH protease binds to the thylakoid membrane and would be physically restricted without swelling and unstacking of the grana thylakoids. Deg proteases are peripherally attached to the membranes with Deg 1, 5, and 8 in the lumen and 2 and 7 on the stromal side. FtsH protease is P dependent but Deg proteases are not. Thylakoid swelling enables Deg 1, 5, and 8 to move freely within the lumen.

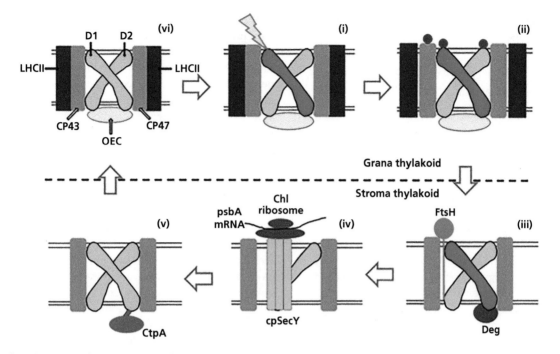

Fig. 6.8 Six stages of damage and repair of PSII in higher plant chloroplasts. Components are labelled for stage vi (upper left): LHCII, light harvesting complex; D1 and D2, transmembrane protein subunits associated with the oxygen evolving centre (OEC); CP43 and CP47, chlorophyll proteins that bind chlorophyll molecules. (i) Light inactivation of PSII and damage of the D1 protein. (ii) Phosphorylation of inactive PSII by phosphokinase, STN8. (iii) Degradation of damaged D1 by proteases FtsH and/or Deg. (iv) Synthesis of new D1 protein by a chloroplast ribosome using psbA mRNA and its simultaneous insertion in PSII by a thylakoid-transmembrane translocan, cpSecY. (v) Processing of D1 by peptidase CtpA. (vi) Return of active PSII to the grana thylakoid. Reproduced from Miyata K. et al. (2012) with permission from Kluwer/International Society of Photosynthesis Research.

Synthesis of D1 protein and its insertion into PSII (Fig. 6.8) requires transcription and translation from chloroplast genes—likely to be under thioredoxin control (Järvi et al. 2015). Puthiyaveetil et al. (2014) suggest that ribosomes, which are comparatively large in size, are confined to stroma lamellae where the new protein is synthesized and that the assembly of PSII occurs in the grana margins and/or stroma lamellae. The reassembled PSII complex returns to the stacked grana. Järvi et al. (2015) list PSII auxiliary proteins with their location in the membrane, lumen, or stroma and discuss, where possible, their function.

It has been proposed that the retention of genes in the chloroplast genome for protein subunits of energy-transducing enzymes facilitates their control by the redox system. This is the CoRR hypothesis: the colocation of gene and gene product for redox regulation of gene expression (Allen 2015).

Degradation of one molecule of photodamaged D1 requires ~240 molecules of ATP and synthesis requires more than 1,300 molecules (Murata and Nishiyama 2017). Provision of this ATP is an important function of CEF. Miyata et al. (2012) calculated the cost and benefit of PSII repair in spinach in relation to plant growth under high light (HL, 300) or low light (LL, 120 μmol m^{-2} s^{-1}) (8 h light/16 h dark) and with damage induced by 2 h of light ranging from 400 to 3,000 μmol m^{-2} s^{-1}. Rates of repair increased as plants received up to 1,600 μmol m^{-2} s^{-1} but decreased above that, which they suggest was probably due to increased ROS production. Generally, the benefits of repair were more than 35–270 times the costs at PPFD from 400 to 1,600

μmol m^{-2} s^{-1} and benefits were greater for HL than LL plants.

6.3.2 Node 2: control of NADPH/ATP ratio, CEF, and redox reactions

The photosynthetic apparatus adjusts to changes in environmental conditions and requirements for ATP and NADPH in a way that avoids an imbalance in production between them and their metabolic consumption which might otherwise lead to disturbance of cellular redox condition, and an increased production in reactive oxygen (Schöttler and Tóth 2014) could lead to destruction of the photosynthesis apparatus. This governor controls functioning of PSI and the PQ pool. Walker et al. (2020) suggest that the energy balancing effects of CEF may adapt to longer-term energy demands through enzyme expression.

There are two operands: the enzymes NADH/ferredoxin-plastoquinone reductase and ferredoxin-plastoquinone reductase. The NADPH/ATP ratio (command) produces an increase/decrease in ATP production (output). Cyclical electron flow is an essential component of the protection systems because its action produces ATP even though LEF is reduced and this ATP may be essential for the repair process of PSII.

NADPH, essential for the CBB cycle, is produced from ferredoxin reduced by electrons excited at PSI. However, reduced ferredoxin is not only involved in NADPH production but also an electron donor in a number of enzymatic reactions and is considered by Oelze et al. (2008) to have a central distributing role in redox regulation of the photosynthesis process (Fig. 6.9), including the production of reduced thioredoxin. Thioredoxins are signalling molecules that occur in the cytosol, mitochondria, and plastids, though the chloroplast has the largest numbers and amounts (Oelze et al. 2008). Thioredoxin reduces disulphides of target proteins and changes the activation state of enzymes. Thioredoxin proteins are also involved in antioxidant defence processes by donating electrons to peroxiredoxins, which reduce peroxides.

Ferredoxin can also donate electrons to the pool of plastoquinone, an essential step in CEF. The CBB cycle uses a ratio 1.5 ATP/NADPH and LEF only produces a ratio of 1.28. A constant proportion of CEF can be expected even in the steady state, estimated at ~15% of the total LEF flux. Additionally when LEF is reduced as a result of PSII damage the ATP generated as a result of CEF may be essential for repair. PSI accepts electrons from plastocyanin (Fig. 6.1) that are further excited by light absorption by chlorophyll to form P700^{+}. Three iron-sulphur proteins known as Fe-S centres are electron acceptors and damage can be caused when supply of electrons exceeds the capacity of electron acceptors.

CEF has two functions: it generates ATP and restricts accumulation of electrons at the PSI acceptors, protecting PSI from high light stress. In CEF, electrons from ferredoxin are donated to plastoquinone by two principal routes (Fisher et al. 2019): the enzymes NADH/ferredoxin-plastoquinone reductase and ferredoxin-plastoquinone reductase (the PGRL1/PGR5 path; Hertle et al. 2013). These enzymes are inhibited by ATP and Fisher et al. propose that CEF is activated when ATP is low but then is downregulated as ATP levels increase. They further suggest that differences in enzyme properties result in the NADPH pathway being activated with moderate decrease in ATP but the ferredoxin pathway is activated under greater ATP depletion.

Roach and Krieger-Liszkay (2012) suggest that a tight coupling between the cytochrome b$_6$f complex and PSI is required for CEF. CEF through the PGR5 pathway also contributes to generation of ΔpH that induces NPQ and so reduces electron flow (Munekage et al. 2002).

It is generally considered that PSII is more susceptible to photoinhibition than PSI (Keren and Krieger-Liszkay 2011). However, a possible process that can lead to damage is the Mehler reaction (Asada 2006) whereby oxygen molecules are reduced resulting in formation of superoxide anion radicals (O$_2^-$). Hydrogen peroxide (H$_2$O$_2$), produced from the superoxide, destroys the Fe-S centres (Sonoike 2011). Furthermore, genetic pathways leading to cell death can be activated, for example by singlet oxygen through a signal transduction sequence (Queval and Foyer 2012).

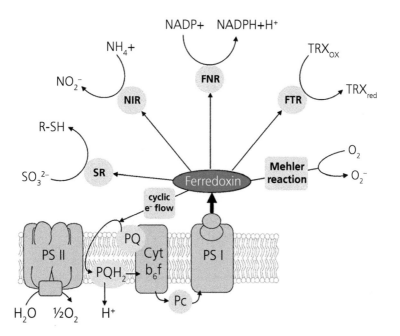

Fig. 6.9 Multiple enzyme reactions in which reduced ferredoxin is an electron donor: FNR (ferredoxin-NADP$^+$ reductase) produces NADPH used in the CBB cycle; FTR (ferredoxin-thioredoxin reductase) produces reduced thioredoxin (TRX$_{red}$) involved in multiple enzyme activations; SR (sulphur reductase); NR (nitrate reductase); the Mehler reaction; and as a donor to the plastoquinone pool (PQ) that initiates CEF resulting in generation of protons, and so ATP, without generating reducing power. Reproduced from Oelze et al. (2008) with permission from Elsevier.

Two scavenging enzymes of ROS are arranged near to the PSI reaction centre: superoxide dismutase (SOD), which converts O_2^- to hydrogen peroxide, and ascorbate peroxide (APX), which converts the hydrogen peroxide to water and so may prevent or restrict damage. Moreno et al. (2008) review light-generated reducing power inside the chloroplast and the roles of ascorbate and glutathione in buffering against potential damage. Glutathione is a major redox buffer. The flux of electrons that are derived from water oxidation by PSII and then produce water at PSI has been termed the water–water cycle (WWC). Cai et al. (2017) provide an example where the WWC is an important electron sink when CO_2 assimilation is restricted due to stomatal closure in *Camellia*.

Multiple enzymes have a peroxidase function and may reduce possibility of damage. Dietz (2016) suggests reasons for this may include: localization, where the enzymes may be linked to different

thylakoid structures; specificity for different peroxide substrates; coupling with different metabolic pathways; and regulation by different environmental and developmental cues.

There are control linkages between the CEF system and both the CBB cycle and PSII. Strand et al. (2017) suggest a relative deficit of ATP would slow the CBB cycle leading to accumulation of reduced NADPH and ferrodoxin that can reduce O_2 to superoxide and H_2O_2, so that H_2O_2 would be a good indicator of imbalances in the production and consumption of ATP/NADPH.

Johnson (2018) proposes that changes in thylakoid stacking regulate the LEF/CEF balance by altering the degree of partition of the PQ pool. The NADPH/ATP ratio in the chloroplast controls the activity of the protein kinase enzyme STN7 that phosphorylates some LHCII proteins, resulting in their detaching from PSII and binding to PSI and also reduction in the sizeof the grana. PQ reduction in the light activates STN7 and its activity is then regulated by the stromal redox state and so the extent of thioredoxin (TRXf) reduction. When the NADPH/ATP ratio is low, thioredoxin is oxidized, allowing ATP binding and phosphorylation of LHCII. Alternatively, when ATP is in deficit (high NADPH/ATP ratio) STN7 is inactivated by reduced thioredoxin, leading to larger grana stacks. The amount of thylakoid membrane stacking and thus the LEF/CEF balance are therefore regulated by the metabolic state of the chloroplast.

6.3.3 Node 3: controls that sustain the CBB cycle

Recall reports that RuBisCo amount is not limiting to the carboxylation reaction (Raines 2003). The inhibition process at RuBisCo reaction sites, which is continuous and sometimes actively controlled, as with 2-carboxy-D-arabinitol 1 phosphate (CA1P), should be considered as coupled with the action of rca in the haloenzyme RuBisCo ← rca in a control subsystem, providing protection against enzyme proteolysis of RuBisCo and modulation of its enzymatic activity in response to the activation of rca by reduced thioredoxin and ADP/ATP ratio. RuBisCo ← rca is the *operand*, thioredoxin and ATP/ADP the *command*, and RuBisCo activity the *output*. Other enzymes in the CBB cycle have a similar control structure and effect.

RuBisCo exists in large quantities in foliage and has been described as the most abundant protein on earth. However, RuBisCo has long been considered an enigma. Although essential for carbon fixation it has been suggested as having a slow reaction rate relative to other enzymes, but this is disputed by Bathellier et al. (2018). It can also catalyze an oxygenase reaction resulting in CO_2 production from RuBP (Chapter 3)—the process of photorespiration (Andersson 2008)—rather than CO_2 fixation.

The first requirement for the carboxylase reaction is carbamylation to activate a reaction site—this involves CO_2 and then association with Mg_2^{2+} (Stec 2012). This CO_2 is not used in the carboxylation of RuBP but once activated the site can fix CO_2 to RuBP, forming a six-carbon phosphorylated molecule that rapidly decays into two molecules of glycerate-3-phosphate (Andersson 2008).

RuBisCo has a low catalytic efficiency (compared with other enzymes) with a carboxylase constant (k_{cat}) of ~3 s^{-1} when calculated assuming Michaelis–Menten kinetics. However, the concentration of RuBisCo within the stroma typically exceeds the concentration of its substrate and Harris and Königer (1997) suggest this could facilitate organized enzyme sequences and metabolic channelling with other enzymes in the CBB that are also present in higher concentrations than their substrates (Harris and Königer 1997, Table 1). CO_2 is mostly stored in cells as bicarbonate and the supply of CO_2 from bicarbonate is limited by activity of carbonic anhydrase. Igamberdiev and Roussel (2012) suggest the flux of CO_2 to RuBisCo is a process limiting CO_2 assimilation, rather than a steady state concentration. The reaction will be saturated at the point when the CO_2 flux approaches its maximum. Even when plants photosynthesize in full sunlight not all of the reaction sites of RuBisCo are active (Raines 2003).

The carboxylation reaction is not an inevitable consequence of RuBP association with a reaction site. The enzymatic process may misfire, leaving the reaction site blocked to carboxylation (Carmo-Silva et al. 2015). Binding of reaction sites can be by a number of sugar phosphate molecules, most notably of course RuBP, and in some species a specific molecule, CA1P, is produced, particularly at night, that inhibits further reactions (Andersson 2008). This blocking has the important function of protecting RuBisCo against proteolysis and loss of catalytic activity by endogenous and exogenous proteases (Khan et al. 1999) and so may prolong the enzyme's existence (Carmo-Silva et al. 2015).

Blocked reaction sites are unblocked by the enzyme rca in the induction process (Robinson and Portis 1989a; Portis et al. 2008; Carmo-Silva and Salvucci 2011) (Chapter 3). rca is an AAA+ enzyme (Atpase Associated with diverse cellular Activities)—enzymes that act as mechanical motors driven by ATP (Erzberger and Berger 2006) in enzymatic processes involving large molecules (e.g. oligomerization), which is what is required to make rca active. rca is composed of four isomers organized by an ATP-requiring process into an active form that identifies with inhibiting molecules at RuBisCo reaction sites (Snider et al. 2008).

Bhat et al. (2017) suggest rca docks onto RuBisCo, one active site at a time, and that the pulling force it exerts on the blocking molecule is tuned to avoid global unfolding of RuBisCo. rca counteracts progressive inhibition, or 'fallover' (Robinson and Portis 1989b; Carmo-Silva et al. 2015), of RuBisCo by sugar phosphates and/or CA1P and has been called RuBisCo's molecular chaperone (Carmo-Silva and Salvucci 2011). Having a switching/amplifying system for reaction sites that can ensure utilization of periodic illumination and prolong the active life of a large molecule seems likely to contribute to the stability of the photosynthesis system.

There are two isoforms of rca (e.g. Ayala-Ochoa et al. 2004; Vargas-Suarez et al. 2004), differing only at a carboxyl terminus, and Zhang and Portis (1999) suggest that in plants containing both isoforms rca activity is regulated by the levels of ADP/ATP and the redox potential via thioredoxin, and this adjusts the rate of CO_2 assimilation to prevailing light flux (Salvucci et al. 1985; Robinson and Portis 1989a). Mott and Woodrow (2000) note that the ratio of rca to RuBisCo was found to be approximately five times greater for understory plants in a rainforest than for tobacco. Moreno et al. (2008) suggest the enzymatic activity may be affected through redox effects on rca.

Oelze et al. (2008) propose redox-dependent signalling is the core of photosynthesis regulation, although it is important to note that this occurs along with a change in quantities of ATP and $NADPH^+$. Oelze et al. suggest that 11 out of 12 enzymes of carbon fixation are affected by redox state. The signalling molecule reduced thioredoxin provides feedforward control which is important when light flux may change rapidly and induction of dark cycle enzymes is necessary. The requirement of enzymes of the CBB cycle for reduced ferrodoxin can protect against metabolite exhaustion when light declines and so maintain capacity to recommence activity without a long time lag.

6.3.3.1 Photorespiration and the C_2 cycle

Although the oxygenase reaction of RuBisCo is frequently considered to be detrimental because it reduces carbon uptake, it may have an important function in maintaining electron flow. When CO_2 supply is reduced as during stomatal closure, the regeneration of ATP used in the oxygenation reaction maintains electron flow (four electrons per oxygen reaction; Busch et al. 2018) and so can contribute to protection of the photosynthesis system against photo-oxidation. This is likely particularly important at higher temperatures (i.e. >27°C) because increase in temperature lowers the concentration of CO_2 in aqueous solution more than that of O_2, so that generally the oxygenase function increases relative to carboxylase as temperature increases (Jordan and Ogren 1984). The oxygenation reaction produces 2-phosphoglycerate which, through a series of reactions, results in the carbon skeleton for nitrate assimilation in foliage.

Photorespiration can be important under conditions that produce stomatal closure and which may lead to depletion of leaf CO_2. Chastain et al. (2014) found in commercial cultivars of *Gossypium hirsutum* (cotton) that in response to water deficit there was decrease in stomatal conductance and, while electron transport did not decrease, there was increase in photorespiration and the ratio of dark respiration to gross photosynthesis—and plants had reduced cotton (lint) yield. They comment that:

> increase in RuBisCo oxygenation reaction, along with the lack of a drought-induced inhibition of photosynthetic electron transport, suggests that decrease[s] in P_G (gross photosynthesis) are primarily due to a reduction in the RuBisCo carboxylation reaction in response to lack of substrate ... rather than a down regulation of RuBisCo.

They suggest, on the basis of leaf temperature, that confounding effects of temperature were unlikely but note that leaves sampled prior to flowering were more sensitive to water deficit than those after flowering. Muraoka et al. (2000) found that midday depression of CO_2 assimilation in sun leaves of *Arisaema heterophyllum* was due to increased photorespiration while that of shade leaves was due to photoinhibition.

The oxygenation reaction of RuBisCo results in the production of one molecule of 3-phosphoglycerate, rather than two molecules as in the carboxylation reaction, and one molecule of 2-phosphoglycerate, plus CO_2. The 2-phosphoglycerate is the starting molecule for the C_2 oxidative photosynthetic cycle (Taiz et al. 2022, section 10.2) which ultimately returns these carbons to the CBB cycle. The C_2 cycle includes reactions occurring in the peroxisome and mitochondrion as well as the chloroplast and involves amino acids in a sequence of reactions that result in formation of 3-phosphoglycerate.

Key reactions of the C_2 oxidative cycle are: formation of glycolate from 2-phosphoglycerate that is transported from the chloroplast and diffuses to the peroxisome; in the peroxisome glycolate is transformed to the amino acid glycine through interaction with glutamate; glycine leaves the peroxisome and enters the mitochondrion, where two molecules act with NAD+ to produce one molecule of NADH, NH4+, which diffuses to the chloroplast and results in formation of glutamate, and one molecule of serine which diffuses to the peroxisome and is transanimated and reduced to form glycerate which then reenters the chloroplast, and is phosphorylated and so can be a substrate in the CBB regeneration reactions.

When reducing conditions in the chloroplast increase under stress then reductants are transported to the cytosol through what is termed the malate/oxaloacetate (OAA) shuttle (the malate valve) (Taniguchi and Miyake 2012). Stromal OAA is reduced to malate by malate dehydrogenase (MDH) using NADPH. Malate is then exported by a transporter and cystolic NAD-MDH produces NADH by oxidizing the exported malate to OAA which is then returned to the chloroplast stroma. NADH reduces NO_3^- to NO_2^- which is the first step in the assimilation of NO_3^-. Rachmilevitch et al. (2004) show that nitrate assimilation in shoots of both a eudicot (*Arabidopsis thaliana*) and a monocotyledon (*Triticum aestivum*) depend upon photorespiration. Busch et al. (2018) show, using fertilizer experiments with *Helianthus annus*, that reduction of NO_3^- to NH_4^+ is associated with photorespiration driven nitrogen assimilation and suggest this 'provides a beneficial sink for excess electrons and helps the ATP:NADPH budget' as per Noctor and Foyer (1998).

6.3.3.2 Regeneration of dark cycle substrates

The CBB cycle comprises 11 enzymes (see Raines 2003) and flux control through the CBB by different enzymes varies according to growing conditions. The CBB cycle involves large fluxes and Bathellier et al. (2018) suggest that 'it has to be tightly regulated to avoid rapid and huge metabolite accumulation and divergence from steady state'. Enzymes of the CBB cycle are activated by thioredoxin so that the dark cycle activity is triggered by onset of LEF. These enzymes are deactivated by oxygen and this deactivation will proceed in the dark when the ferredoxin → thioredoxin reduction stops. Flux control can be determined by different enzymes in the cycle. While RuBisCo limits CO_2 fixation under conditions where photorespiration may be favoured, particularly high light flux, under other conditions enzymes controlling regeneration leading to increased RuBP may be more important.

The use of transgenic plants in which levels of CBB cycle enzymes are manipulated has shown that under a range of environmental conditions the level of RuBisCo protein has little impact on the control of carbon fixation (Raines 2003.) Raines suggests: 'The balance between regeneration and carboxylation capacities is normally maintained such that they co-limit photosynthesis and neither of these capacities dominate control of carbon dioxide in the Calvin cycle.'

Considering the CBB cycle Morandini (2009) suggests: 'The fact that most reactions far from equilibrium are subject to regulation suggests that regulations are important to prevent the reactions continuing to convert substrates into products. Otherwise the 3PGA pool will enlarge enormously engulfing most of the chloroplast carbon and Pi, certainly a dangerous situation.' Metabolic regulation controls concentration not flux and this metabolite is important to avoid the constraints of moiety concentration.

6.3.4 Gas exchange studies

Control by quantities whether light → (ATP + NADPH) and/or CO_2 uptake via stomata is always present; a curvilinear A/Q relationship can always be found but its parameters may vary with change of conditions for any one foliage element (Chapter 3). Study of photosynthesis through analysis of gaseous exchange has focussed on the relative importance of stomatal conductance (CO_2 diffusional limitation) and RuBisCo carboxylation (electron transport limitation) (Flexas and Diaz-Espejo 2015). These two processes are the basis of the widely used Farquhar et al. (1980) model which defines the 'teeter-toter' or 'see-saw' (Farquhar et al. 2001) between these two rate limitations as they are affected by physiological processes and environmental conditions. An important assumption of this model is that the substomatal CO_2 concentration (C_i) represents the CO_2 concentration at the carboxylation site inside chloroplasts so that the mesophyll conductance (g_m) of CO_2 between the substomatal cavity and the chloroplast is infinite.

Flexas and Diaz-Espejo (2015) point to a substantial amount of research that now suggests that g_m is finite and variable and is a third limiting factor to be considered in gaseous exchange studies. Analysis of the physiological processes contributing to g_m could enable gas exchange studies and models to be integrated more effectively with photosynthesis physiology.

Three features influencing gaseous transport may affect g_m: cell wall properties, cell membrane properties, and chloroplast morphology and distribution (Flexas and Diaz-Espejo 2015). Veromann-Jürgenson et al. (2017) found that low g_m was the main limiting factor of photosynthesis in 11 gymnosperm species and suggest the strongest sources of limitation were extremely thick mesophyll cell walls and high chloroplast thickness and low exposed surface area of chloroplasts (S_c). Flexas and Diaz-Espejo (2015) suggest that chloroplast surface area facing the intercellular air space may change with temperature and that this may explain a decline in g_m with increasing temperature observed by von Caemmerer and Evans (2015). Peguero-Pina et al. (2017) found g_m the most limiting factor for photosynthesis per unit leaf area within provenances of the Mediterranean sclerophyll *Quercus ilex* and attributed these differences in part to cell wall thickness and S_c exposed to intercellular air space, but obviously processes outlined in this chapter would also affect g_m. However, gaseous exchange studies would need to be studied under a wide range of conditions to analyse this.

6.4 Conclusion

The three reaction nodes essential for photosynthesis, PSII, PSI, and RuBisCo, are each embedded in controls that protect, preserve, or restore their functions in response to changes in light flux. These controls comprise a dynamic system that ensures a degree of stability of the photosynthesis system itself.

Stability of the photosynthesis system is its capacity to maintain competence for CO_2 fixation while receiving fluctuating light flux.

Although there are multiple controls of the photosystem we cannot say that any of them are redundant. Different controls may be accentuated when plants develop under different conditions, as in the case of the repeated predominance of photoinhibition in shade leaves but photorespiration in sun leaves.

The photosynthesis system does not have feedback control in the sense typically used by engineers in the design of homeostatic systems, i.e. with a set point and a monitor of deviation that triggers corrective action in a closed-loop system (e.g. Fig. 4.7). Rather, the controls around each node act as regulators not to optimize the rate of CO_2 fixation but to protect, preserve, or restore the system itself.

6.6 References

Albertsson, P.-Å. 2001. A quantitative model of the domain structure of the photosynthetic membrane. *Trends in Plant Science*, 6, 349–354.

Allen, J. F. 2015. Why chloroplasts and mitochondria retain their own genomes and genetic systems: colocation for redox regulation of gene expression. *Proceedings of the National Academy of Sciences of the United States of America*, 112, 10231–10238.

Andersson, I. 2008. Catalysis and regulation in Rubisco. *Journal of Experimental Botany*, 59, 1555–1568.

Asada, K. 2006. Production and scavenging of reactive oxygen species in chloroplasts and their functions. *Plant Physiology*, 141, 391–396.

Ayala-Ochoa, A., Vargas-Suarez, M., Loza-Tavera, H., Leon, P., Jimenez-Garcia, L. F. and Sanchez-De-Jimenez, E. 2004. In maize, two distinct ribulose 1,5-bisphosphate carboxylase/oxygenase activase transcripts have different day/night patterns of expression. *Biochimie*, 86, 439–449.

Bathellier, C., Tcherkez, G., Lorimer, G. H. and Farquhar, G. D. 2018. Rubisco is not really so bad. *Plant Cell and Environment*, 41, 705–716.

Bhat, J. Y., Miličić, G., Thieulin-Pardo, G., et al. 2017. Mechanism of enzyme repair by the AAA⁺ chaperone Rubisco activase. *Molecular Cell*, 67, 1–13.

Busch, F. A., Sage, R. F. and Farquhar, G. D. 2018. Plants increase CO_2 uptake by assimilating nitrogen via the photorespiratory pathway. *Nature Plants*, 4, 46–54.

Cai, Y. F., Yang, Q. Y., Li, S. F., Wang, J. H. and Huang, W. 2017. The water–water cycle is a major electron sink in *Camellia* species when CO_2 assimilation is restricted. *Journal of Photochemistry and Photobiology. B, Biology*, 168, 59–66.

Carmo-Silva, A. E. and Salvucci, M. E. 2011. The activity of Rubisco's molecular chaperone, Rubisco activase, in leaf extracts. *Photosynthesis Research*, 108, 143–155.

Carmo-Silva, E., Scales, J. C., Madgwick, P. J. and Parry, M. A. J. 2015. Optimizing Rubisco and its regulation for greater resource use efficiency. *Plant, Cell and Environment*, 38, 1817–1832.

Chastain, D. R., Snider, J. L., Collins, G. D., Perry, C. D., Whitaker, J. and Byrd, S. A. 2014. Water deficit in field-grown *Gossypium hirsutum* primarily limits net photosynthesis by decreasing stomatal conductance, increasing photorespiration, and increasing the ratio of dark respiration to gross photosynthesis. *Journal of Plant Physiology*, 171, 1576–1585.

Dekker, J. P. and Boekema, E. J. 2005. Supramolecular organization of thylakoid membrane proteins in green plants. *Biochimica et Biophysica Acta-Bioenergetics*, 1706, 12–39.

Dietz, K. J. 2016. Thiol-based peroxidases and ascorbate peroxidases: why plants rely on multiple peroxidase systems in the photosynthesizing chloroplast? *Molecules and Cells*, 39, 20–25.

Erzberger, J. and Berger, J. 2006. Evolutionary relationships and structural mechanisms of AA+ proteins. *Annual Review of Biophysics and Biomolecular Structure*, 35, 93–114.

Farquhar, G. D., Caemmerer, S. V. and Berry, J. A. 1980. A biochemical-model of photosynthetic CO_2 assimilation in leaves of C3 species. *Planta*, 149, 78–90.

Farquhar, G. D., von Caemmerer, S. and Berry, J. A. 2001. Models of photosynthesis. *Plant Physiology*, 125, 42–45.

Fisher, N., Bricker, T. M. and Kramer, D. M. 2019. Regulation of photosynthetic cyclic electron flow pathways by adenylate status in higher plant chloroplasts. *Biochimica et Biophysica Acta-Bioenergetics*, 1860, 148081.

Flexas, J. and Diaz-Espejo, A. 2015. Interspecific differences in temperature response of mesophyll conductance: food for thought on its origin and regulation. *Plant, Cell and Environment*, 38, 625–628.

Formaggio, E., Cinque, G. and Bassi, R. 2001. Functional architecture of the major light-harvesting complex from higher plants. *Journal of Molecular Biology*, 314, 1157–1166.

Harris, G. C. and Königer, M. 1997. The 'high' concentrations of enzymes within the chloroplast. *Photosynthesis Research*, 54, 5–23.

Hertle, A. P., Blunder, T., Wunder, T., et al. 2013. PGRL1 is the elusive ferredoxin-plastoquinone reductase in photosynthetic cyclic electron flow. *Molecular Cell*, 49, 511–523.

Horton, P., A. V. Ruban, A. V. and Walters, R. G. 1996. Regulation of light harvesting in green plants. *Annual Review of Plant Physiology and Plant Molecular Biology*, 47, 655–684.

Igamberdiev, A. U. and Roussel, M. R. 2012. Feedforward non-Michaelis–Menten mechanism for CO_2 uptake by Rubisco: contribution of carbonic anhydrases and photorespiration to optimization of photosynthetic carbon assimilation. *BioSystems*, 107, 158–166.

Järvi, S., Suorsa, M. and Aro, E. M. 2015. Photosystem II repair in plant chloroplasts—regulation, assisting proteins and shared components with photosystem II biogenesis. *Biochimica et Biophysica Acta-Bioenergetics*, 1847, 900–909.

Johnson, M. P. 2018. Metabolic regulation of photosynthetic membrane structure tunes electron transfer function. *Biochemical Journal*, 475, 1225–1233.

Jordan, D. and Ogren, W. 1984. The CO_2/O_2 specificity of Ribulose 1,5-biphosphate carboxylase oxygenase—dependence on ribulose biphosphate concentration, pH and temperature. *Planta*, 161, 308–313.

Kereïche, S., Kiss, A. Z., Kouřil, R, Boekema, E. J. and Horton, P. 2010. The PsbS protein controls the macro-organisation of photosystem II complexes in the grana membranes of higher plant chloroplasts. *FEBS Letters*, 584, 759–764

Keren, N. and Krieger-Liszkay, A. 2011. Photoinhibition: molecular mechanisms and physiological significance. *Physiologia Plantarum*, 142, 1–5.

Khan, S., John, A. P., Lea, P. J. and Parry Martin, M. A. 1999. 2′-Carboxy-D-arabitinol 1-phosphate protects ribulose 1,5-bisphosphate carboxylase/oxygenase against proteolytic breakdown. *European Journal of Biochemistry*, 266, 840–847.

Kirchhoff, H. 2013. Architectural switches in plant thylakoid membranes. Photosynthesis Research, 116, 481–487.

Kirchhoff, H. 2014. Diffusion of molecules and macromolecules in thylakoid membranes. *Biochimica et Biophysica Acta-Bioenergetics*, 1837, 495–502.

Kirchhoff, H., Hall, C., Woodrow, M. and Reich, Z. 2011. Dynamic control of protein diffusion within the granal thylakoid lumen. *Proceedings of the National Academy of Sciences of the United States of America*, 108, 20248–20253.

Kouřil, R., Oostergetel, G. T. and Boekema, E. J. 2011. Fine structure of granal thylakoid membrane organization using cryo electron tomography. *Biochimica et Biophysica Acta (BBA)—Bioenergetics*, 1807, 368–374.

Latowski, D., Kuczyńska, P. and Strzałka, K. 2011. Xanthophyll cycle—a mechanism protecting plants against oxidative stress. *Redox Report*, 16, 78–90.

Miyata, K., Noguchi, K. and Terashima, I. 2012. Cost and benefit of the repair of photodamaged photosystem II in spinach leaves: roles of acclimation to growth light. *Photosynthesis Research*, 113, 165–180.

Morandini, P. 2009. Rethinking metabolic control. *Plant Science*, 176, 441–451.

Moreno, J., García-Murria, M. J. and Marín-Navarro, J. 2008. Redox modulation of Rubisco conformation and activity through its cysteine residues. *Journal of Experimental Botany*, 59, 1605–1614.

Mott, K. A. and Woodrow, I. E. 2000. Modelling the role of Rubisco activase in limiting non-steady-state photosynthesis. *Journal of Experimental Botany*, 51, 399–406.

Munekage, Y., Hojo, M., Meurer, J., Endo, T., Tasaka, M. and Shikanai, T. 2002. PGR5 is involved in cyclic electron flow around Photosystem I and is essential for photoprotection in *Arabidopsis. Cell*, 110, 361–371.

Muraoka, H., Tang, Y. H., Terashima, I., Koizumi, H. and Washitani, I. 2000. Contributions of diffusional limitation, photoinhibition and photorespiration to midday depression of photosynthesis in *Arisaema heterophyllum* in natural high light. *Plant Cell and Environment*, 23, 235–250.

Murata, N. and Nishiyama, Y. 2017. ATP is a driving force in the repair of Photosystem II during photoinhibition. *Plant Cell and Environment*, 41, 285–299.

Murata, N., Allakhverdiev, S. I. and Nishiyama, Y. 2012. The mechanism of photoinhibition in vivo: re-evaluation of the roles of catalase, α-tocopherol, non-photochemical quenching, and electron transport. *Biochimica et Biophysica Acta (BBA)—Bioenergetics*, 1817, 1127–1133.

Noctor, G. and Foyer, C. H. 1998. A re-evaluation of the ATP:NADPH budget during C_3 photosynthesis: a contribution from nitrate assimilation and its associated respiratory activity? *Journal of Experimental Botany*, 49, 1895–1908.

Oelze, M.-L., Kandlbinder, A. and Dietz, K.-J. 2008. Redox regulation and overreduction control in the photosynthesizing cell: complexity in redox regulatory networks. *Biochimica et Biophysica Acta—General Subjects*, 1780, 1261–1272.

Oguchi, R., Terashima, I., Kou, J. and Chow, W. S. 2011. Operation of dual mechanisms that both lead to photoinactivation of Photosystem II in leaves by visible light. *Physiologia Plantarum*, 142, 47–55.

Peguero-Pina, J. J., Siso, S., Flexas, J., et al. 2017. Coordinated modifications in mesophyll conductance, photosynthetic potentials and leaf nitrogen contribute to explain the large variation in foliage net assimilation rates across *Quercus ilex* provenances. *Tree Physiology*, 37, 1084–1094.

Pinnola, A. and Bassi, R. 2018. Molecular mechanisms involved in plant photoprotection. *Biochemical Society Transactions*, 46, 467–482.

Portis, J. A. R., Li, C., Wang, D. and Salvucci, M. E. 2008. Regulation of Rubisco activase and its interaction with Rubisco. *Journal of Experimental Botany*, 59, 1597–1604.

Puthiyaveetil, S., Tsabari, O., Lowry, T., et al. 2014. Compartmentalization of the protein repair machinery in photosynthetic membranes. *Proceedings of the National Academy of Sciences of the United States of America*, 111, 15839–15844.

Queval, G. and Foyer, C. H. 2012. Redox regulation of photosynthetic gene expression. *Philosophical Transactions of the Royal Society B: Biological Sciences*, 367, 3475–3485.

Rachmilevitch, S., Cousins, A. B. and Bloom, A. J. 2004. Nitrate assimilation in plant shoots depends on photorespiration. *Proceedings of the National Academy of Sciences of the United States of America*, 101, 11506–11510.

Raines, C. A. 2003. The Calvin cycle revisited. *Photosynthesis Research*, 75, 1–10.

Roach, T. and Krieger-Liszkay, A. 2012. The role of the PsbS protein in the protection of photosystems I and II against high light in *Arabidopsis thaliana. Biochimica et Biophysica Acta (BBA)—Bioenergetics*, 1817, 2158–2165.

Robinson, S. and Portis, A. R. 1989a. Ribulose-1, 5-bisphosphate carboxylase/oxygenase activase protein prevents the in vitro decline in activity of ribulose-1, 5-bisphosphate carboxylase/oxygenase. *Plant Physiology*, 90, 968–971.

Robinson, S. P. and Portis, A. R. 1989b. Adenosine triphosphate hydrolysis by purified Rubisco activase. *Archives of Biochemistry and Biophysics*, 268, 93–99.

Ruban, A. V. 2016. Nonphotochemical chlorophyll fluorescence quenching: mechanism and effectiveness in protecting plants from photodamage. *Plant Physiology*, 170, 1903–1916.

Ruban, A. V. 2017. Quantifying the efficiency of photoprotection. *Philosophical Transactions of the Royal Society B: Biological Sciences*, 372, 20160393.

Sacharz, J., Giovagnetti, V., Ungerer, P., Mastroianni, G. and Ruban, A. V. 2017. The xanthophyll cycle affects reversible interactions between PsbS and light-harvesting complex II to control non-photochemical quenching. *Nature Plants*, 3, 16225.

Salvucci, M. E., Portis, A. R. and Ogren, E. 1985. A soluble chloroplast protein catalyses ribulose-biphosphate carboxylase/oxygenase activation *in vivo*. *Photosynthesis Research*, 7, 193–201.

Schöttler, M. A. and Tóth, S. Z. 2014. Photosynthetic complex stoichiometry dynamics in higher plants: environmental acclimation and photosynthetic flux control. *Frontiers in Plant Science*, 5, 188.

Shirao, M., Kuroki, S., Kaneko, K., et al. 2013. Gymnosperms have increased capacity for electron leakage to oxygen (Mehler and PTOX reactions) in photosynthesis compared with angiosperms. *Plant and Cell Physiology*, 54, 1152–1163.

Snider, J., Thibault, G. and Houry, W. A. 2008. The AAA+ superfamily of functionally diverse proteins. *Genome Biology*, 9, 216.

Sonoike, K. 2011. Photoinhibition of Photosystem I. *Physiologia Plantarum*, 142, 56–64.

Stec, B. 2012. Structural mechanism of RuBisCO activation by carbamylation of the active site lysine. *Proceedings of the National Academy of Sciences of the United States of America*, 109, 18785–18790.

Strand, D. D., Livingston, A. K., Satoh-Cruz, M., et al. 2017. Defects in the expression of chloroplast proteins leads to H_2O_2 accumulation and activation of cyclic electron flow around Photosystem I. *Frontiers in Plant Science*, 7, 2073.

Su, X., Ma, J., Wei, X., et al. 2017. Structure and assembly mechanism of plant $C_2S_2M_2$-type PSII–LHCII supercomplex. *Science*, 357, 815–821.

Sundby, C., McCaffery, S. and Anderson, J. 1993. Turnover of the Photosystem II D1 protein in higher plants under photoinhibitory and nonphotoinhibitory irradiance. *Journal of Biological Chemistry*, 268, 25476–25482.

Taiz, L., Zeiger, E., Møller, I. M. and Murphy, A. 2022. *Plant Physiology and Development*, Seventh Edition. Oxford, Oxford University Press.

Taniguchi, M. and Miyake, H. 2012. Redox-shuttling between chloroplast and cytosol: integration of intra-chloroplast and extra-chloroplast metabolism. *Current Opinion in Plant Biology*, 15, 252–260.

van Bezouwen, L. S., Caffarri, S., Kale, R. S., et al. 2017. Subunit and chlorophyll organization of the plant Photosystem II supercomplex. *Nature Plants*, 3, 17080.

Vargas-Suarez, M., Ayala-Ochoa, A., Lozano-Franco, J., et al. 2004. Rubisco activase chaperone activity is regulated by a post-translational mechanism in maize leaves. *Journal of Experimental Botany*, 55, 2533–2539.

Vass, I. 2012. Molecular mechanisms of photodamage in the Photosystem II complex. *Biochimica et Biophysica Acta-Bioenergetics*, 1817, 209–217.

Veromann-Jürgenson, L.-L., Tosens, T., Laanisto, L., and Niinemets, Ü. 2017. Extremely thick cell walls and low mesophyll conductance: welcome to the world of ancient living! *Journal of Experimental Botany*, 68, 1639–1653.

von Caemmerer, S. and Evans, J. R. 2015. Temperature responses of mesophyll conductance differ greatly between species. *Plant, Cell and Environment*, 38, 629–637.

Walker, B. J., Kramer, D. M., Fisher, N. and Fu, X. Y. 2020. Flexibility in the energy balancing network of photosynthesis enables safe operation under changing environmental conditions. *Plants (Basel)*, 9, 301.

Yoshioka-Nishimura, M. 2016. Close relationships between the PSII repair cycle and thylakoid membrane dynamics. *Plant and Cell Physiology*, 57, 1115–1122.

Zavafer, A., Chow, W. S. and Cheah, M. H. 2015. The action spectrum of Photosystem II photoinactivation in visible light. *Journal of Photochemistry and Photobiology. B, Biology*, 152, Part B, 247–260.

Zhang, N. and Portis, A. R. 1999. Mechanism of light regulation of Rubisco: a specific role for the larger Rubisco activase isoform involving reductive activation by thioredoxin-f. *Proceedings of the National Academy of Sciences of the United States of America*, 96, 9438–9443.

PART III

The dynamic relationship between architecture and growth

Introduction to Part III

Plant growth is fundamentally a spatial process. The morphology of a plant combined with the quantity and distribution of foliage and roots determines how space is occupied, resources acquired, and the outcomes of competition decided. Foliage and branches are initiated with a basic spatial regularity, but there are many types of morphology that express this regularity in different structures.

Two important questions about the role of plant morphology in controlling growth are considered:

1. How does genetic variation in morphology affect growth? Variation occurs between and within different taxonomic groupings, but there is substantial evidence of parallel evolution—similar plant structures have evolved in different taxonomic groupings, suggesting some common requirements and function between them.
2. As plants grow, do structural changes occur that may affect the dynamics of growth? This is a particularly important question when considering growth in foliage canopies when there may be competition between individual plants.

Hallé et al. (1978) advanced the theory that plant species grow according to architectural models and define 23 models, describe essential morphological and developmental processes of each, and provide a dichotomous classification system based upon these processes. These models are integrations of botanical components, with a particular emphasis on types of branching—but they are not expressed mathematically. The dichotomous classification structure suggests coherence between the morphological and developmental processes chosen by Hallé et al.—that is, there are functional links between them in how they may affect growth. A strength of their theory is that individual models apply to multiple species, frequently of different taxonomic lineages, which implies that particular architectural models have evolved repeatedly.

In Chapter 7 the theory advanced by Hallé et al. is discussed and illustrates that two additional requirements are necessary to specify the control architecture has on growth processes: first, that foliage amount and distribution can vary markedly between species that have the same branching pattern and should be included in definitions of architecture.

Second, plasticity should be an integral part of models. Growth within foliage canopies (Chapter 8) involves two types of plasticity: resource acquisition plasticity in response to light quantity and which can result in change in shape of crowns, and morphogenetic plasticity in response to changes in light quality found in canopies. These processes are determined by the genetics of plants. An example is given of change productivity of a foliage canopy that involves manipulation of the dynamics between the environment and plasticity.

An understanding of how plant growth is controlled requires integration of knowledge about morphological development, including plasticity, with that of physiology.

Reference

Hallé, F., Oldeman, R. A. A. and Tomlinson, P. B. 1978. *Tropical Trees and Forests: An Architectural Analysis.* Berlin, Springer.

CHAPTER 7

Variation in architectural dynamics and its effects on growth

7.1 Plant architecture

Hallé et al. (1978) present a theory of plant architecture defined by morphological features and growth processes. The objective of the theory is to account for how variations in plant form may affect the dynamics of growth in tropical trees. They suggest the long time that tropical forests have been in existence has provided an environment for sustained evolution, that the resulting species diversity provides a wide variety of plant architectures, and the study of tropical trees can be used to identify generally important components of architecture in growth. The classification of architectures they present transcends taxonomic classifications.

7.1.1 Classification of architectures

Hallé et al. define architecture as: 'The visible, morphological expression of the genetic blueprint of a tree at anytime.' They recognize 23 architectural models based on the lifespan of meristems and differentiation of vegetative meristems. Their intent with an *architectural model* is to define successive architectural phases and their effects on the growth process. Chomicki et al. (2018) give examples of herbaceous species conforming to 16 of the models defined by Hallé et al.

The main criteria for identifying models are concerned with extension growth, the single' most important being lifespan—for example whether a meristem produces monopodial (resulting in a single trunk) or sympodial growth. Hallé et al. give four examples of criteria for which there are alternative forms:

1. Sexual (determinate) as opposed to vegetative (indeterminate) differentiation. *Zea mays* is an example where differentiation of an apex to sexual development terminates vegetative growth (Chapter 2).
2. Orthotropy (axes that are erect, commonly with spiral or decussate phyllotaxis; Fig. 7.1a) as opposed to plagiotropy (axes that are horizontal and dorsiventrally symmetrical; Fig. 7.1b). Hallé et al. consider this division of architectural importance: plagiotropic branches are typically found in species occurring in shaded habitats, while orthotropic branches are typically in species from more open habitats.
3. Rhythmic growth (episodic) (Fig. 2.2) can result in marked segmentation of the plant body, as opposed to continuous growth (Fig. 7.1c). An indication of continuous growth can be absence of a pronounced morphological segmentation of the shoot. Plants with continuous growth tend to have internodes, foliage, and lateral meristems of the same size; see example of *Gossypium hirsutum* in Chapter 8.
4. Differences in the chronology of branch development can occur such as late specialization of one axis among a group of contemporaries initially all alike. For example, delayed growth of a shoot meristem is the characteristic of proleptic growth: *Pinus longaeva* provides an example (Chapter 2) in response to damage, but some plants have delayed growth as a normal feature (Chapter 10).

Hallé et al. present a dichotomous key for classifying trees into the 23 architectural models and identify multiple species following each model.

The Dynamics of Plant Growth. E. David Ford, Oxford University Press. © E. David Ford (2023). DOI: 10.1093/oso/9780192867179.003.0007

Fig. 7.1 Illustrations of criteria used by Hallé et al. in the classification of plant architectures. (a) Orthotropic branching: branches from the main axis, the trunk, slope upwards. As the tree grows and a branch position becomes lower in the crown, the older part of the branch, i.e. that next to the trunk, becomes more horizontal but its new shoots still slope upwards. Illustration of *Pinus sylvestris*. (b) Plagiotropic branching: branches from the main axis are generally horizontal. Illustration of *Araucaria columnaris* with a developed whorl of four branches and a new whorl forming at the apex of the main axis. (c) Continuous growth: illustration of *Eucalyptus globulus*. In contrast to rhythmic growth there is no endogenously controlled pause and restart typically associated with development of a segmented architecture, although growth rates of axes can vary and even pause. All axes have an equivalent structure. Photograph credits: (a) Mikkel Eye https://meye.dk; (b) iStock; (c) Royal Botanic Gardens Kew, licensing@kew.org.

They stress that these models, which they name after botanists who have typically researched representative species, are particular points in a continuum of possible models. An essential feature is that the same architectural model applies to species that are not taxonomically close, which they suggest indicates some 'architectural structural convergence' and so models 'represent(s) a biological necessity for functional and competitive disposition and growth of organs in plants' under a particular set of conditions. That models should provide causal explanations for the dependence of growth of a number of different species on a particular architecture depends upon the process of *convergent evolution* (Box Fig. 7.1). There has been independent evolution of phyllotaxis in the major plant lineages and Véron et al. (2021) present evidence and reason that 'similar molecular regulatory modules have been deployed repeatedly ... and drive apical function in convergent shoot forms'.

Rauh's, Massart's, and Attim's models are typical of trees with multiple branches arising from a single orthotropic trunk and provide an example of how architectural differences may contribute to growth in different environments (Fig. 7.2). They differ in whether the branches are established as orthotropic or plagiotropic and whether growth is rhythmic or continuous.

Rauh's model is one of the most frequent among seed plants and Hallé et al. considered it as a base case with branches that are initiated orthotropically, all shoot apical meristems are equivalent and rhythmic, and branches occur in tiers. Flowers are always lateral. Hallé et al. suggest that the ability of trees of Rauh's model to regenerate an existing crown is important: 'if the trunk meristem is destroyed it is readily replaced, usually by the uppermost lateral meristem or, if the damage is more extensive, by the uppermost branch which rapidly substitutes as a leader'. Most species of *Pinus* have this model, e.g. *Pinus sylvestris* (Fig. 7.2a), and see also Fig. 7.1a, but it is common throughout Eudicots, e.g. *Euphorbia* and *Quercus* species, and occurs in some shrubs, e.g. *Calluna vulgaris* (Hallé et al. 1978, pp.226–227). Clearly, with a widely occurring model the question arises whether differences in growth between species classified within the same model might be influenced by additional architectural features that have not been considered. This is discussed later in section 7.2.1, in application of Rauh's model to *Picea* and variation in growth between individuals of the same species.

In Massart's model the architecture also comprises 'an orthotropic monopodial trunk with rhythmic growth and which consequently produces regular tiers of branches at levels established by the growth of the trunk meristem. However, branches

Box 7.1 Convergent evolution

Convergent evolution is the separate development of the same or similar features, such as a particular feature of plant architecture, in taxonomically unrelated or distantly related species but that exist in similar habitats.

Leverenz et al. (2000) provide an example of the development of shade shoot architecture in conifers native to New Zealand (NZ) compared to northern hemisphere (NH)

conifers (Leverenz 1996). Under shade conditions foliage arrangement can become spread from supporting shoots with less overlapping foliage elements than found in full light. This spread can be measured using the ratio of a shoot's silhouette area to the sum of the area of all foliage when removed from the shoot—shoots that have no overlap in foliage would have a value of 1. The maximum found

Box Figure 7.1 (a) NH Pinaceæ: *Picea sitchensis*. Sun foliage with needles forming a cylinder around the shoot. (b) *P. sitchensis* shade foliage partially flattened in two parallel rows on either side of the shoot, R_{max} 0.74. (c) NH *Abies grandis*, generally considered as a shade-tolerant species with flattened shoot and branch system and foliage in two ranks along the shoot, R_{max} 0.99. (d) NZ Podocarpacæ: *Prumnopitys ferrugine*. Similar foliage element size 41 mm^2 and structural arrangement to that of *Abies grandis*, R_{max} 0.97. (e) NZ Araucariaceæ. *Agathis australis* (kauri pine). Foliage element size 289 mm^2, R_{max} 0.96. (f) NZ Podocarpacæ. *Phyllocladus totara*. Foliage element size 60 mm^2, R_{max} 0.85. Photograph credits: (a) Tree Seed Online, https://www.treeseedonline.com/; (b) https://lh2treeid.blogspot.com/; (c) Wikimedia commons; (d) University of Auckland, New Zealand, https://www.nzplants.auckland.ac.nz/; (e) Oregon State University, https://landscapeplants.oregonstate.edu; (f) University of Auckland, New Zealand, https://www.nzplants.auckland.ac.nz/.

Box 7.1 *Continued*

for this ratio, termed R_{max}, in shade foliage of a species has been considered as an indicator of shade tolerance in NH conifers.

Conifers in NZ and the NH have been geographically isolated for ~150 My (million years). Foliage size of NZ conifers covers a wider range than that found in NH, but Leverenz et al. found the range of R_{max} to be similar and conclude: 'It appears that natural selection has resulted in species with R_{max} approaching the maximum possible value of 1.0 in both NZ and NH conifers.'

Fig. 7.2 Three images of plants conforming to particular architectural models and with diagrammatic representations of their models below the images showing increase in size from the right. (a) *Pinus sylvestris*, Rauh; (b) *Araucaria columnaris*, Massart; (c) *Eucalyptus globulus* Attim. Photograph credits: (a) Canstock, Serge64; (b) Pinterest; (c) FAVPNG. Line diagrams reproduced from Hallé et al. (1978) with permission from Springer.

are plagiotropic, either by leaf arrangement or symmetry' (Hallé et al. 1978, p.191). *Araucaria columnaris* (Fig. 7.2b) is an example and which develops precise crown symmetry, and see also Fig. 7.1b. Many species follow this model, mainly forest trees, growing at different levels in forest canopies. Hallé et al. suggest that the plagiotropic organization of branches 'confers a high individual survival potential . . . in the lower stories of the forest since light interception is efficient'. Additional examples are found in the gymnosperms, e.g. *Abies alba* and *Abies amabilis*, both of which are shade-tolerant species, as well as members of the persimmon/ebony genus *Diospyros*, e.g. *D. discolour*, the velvet persimmon. The role of architecture in shade tolerance is discussed further in Chapter 8.

Attim's model is also similar to that of Rauh in having an orthotropic trunk but, as distinct from

Massart's model, orthotropic branching. However, it has continuous rather than rhythmic growth, which results in what Hallé et al. term diffuse branching, e.g. not arranged clearly in tiers (Fig. 7.2c, *E. globulus*). Hallé et al. point to their being fewer examples of Attim's model compared to that of Rauh and suggest this may be due to lack of a process for suspending growth during less favourable seasons so that this type of growth may be restricted to stable environments. However, Hallé et al. do suggest that fluctuation in growth occurs in plants with Attim's model growing in seasonal climates but that it is not endogenous. The comparative lack of regularity in morphology in species following Attim's model may enable branching that appears to explore its environment—discussed with the example of *Thuja plicata* in Chapter 8.

An important contribution of Hallé et al. is their proposed set of widely applicable models for understanding how plant architecture functions in the growth process. But this raises the question of the extent to which particular model features, or their combinations, may control growth: some understanding of this can be obtained from study of the evolution of architecture.

7.1.2 Evolution of conifer architectures

Leslie et al. (2012) define lineages within the Coniferales and present a dated phylogeny of 489 conifer species representing some 80% of the diversity. The phylogeny is based on a molecular data set (Box Fig. 7.2) using two nuclear genes and two chloroplast genes and 16 fossil calibration points. Leslie et al. (2018) comment further on fossil dating of different conifer lineages.

Box 7.2 Dating phylogenies

Molecular data, particularly from DNA, are used to construct phylogenies and estimate the time when divergences in the phylogeny have occurred, called molecular dating. Sauquet (2013) provides a practical guide. Paleontological dating is limited by the occurrence of fossils and molecular dating can provide more complete relative divergence times but needs to be calibrated using the fossil record to estimate absolute times.

Typically, one, or a few, genes are selected and nucleotide sequences within a section of a gene are determined for all plants in the study. Relationships between these individuals are determined according to similarities and differences found in the sequences. Computer analyses are used to construct groupings for the individuals in the study with programmes implementing statistical approaches for defining groupings.

Molecular dating of divergence between lineages in a phylogeny depends on the assumption of a molecular clock: that there is a regular change over time in the composition of large biomolecules such as DNA, RNA, or amino acid sequences for proteins. Most changes are neutral, i.e. mutations with no effect on fitness. However, the rates of change can vary between types of organism and over time. This has led to the use of statistical techniques in the calculation of relaxed molecular clocks.

Conifers have an extensive and well-documented fossil record and are of interest in having existed since the Permian (299–252 My before present) and so evolved through substantial continental drift and periods of marked climate change. Leslie et al. were particularly interested in comparing evolution between northern and southern hemispheres which have different distributions of land mass that could have affected biological diversity. Leslie et al. report: 'The dated phylogeny shows that most extant conifer species diverged in the Neogene but belong to lineages that generally diverged earlier in the Mesozoic' (Neogene 23–2.6 My before present; Mesozoic 252–66 My before present). Extant refers to existing living species.

The phylogeny is represented as a tree extending from the central point of a circle (Fig. 7.3) and radiating out to major clades: for the northern hemisphere, Taxaceæ, Cupressaceæ, Pinaceæ, and the Sciadopityaceæ which has a single species, the Japanese umbrella pine *Sciadopitys verticillata*; for the southern hemisphere, Araucariaceæ and Podocarpaceæ. Early divergence points, those between Sciadopityaceæ and Taxaceæ/Cupressaceæ and between Araucariaceæ and Podocarpaceæ, are dated at close to 251 My, i.e. within the Permian. Divergencies between these clades occurred before the Mesozoic–Cenozoic boundary at 65 My and all clades survived the

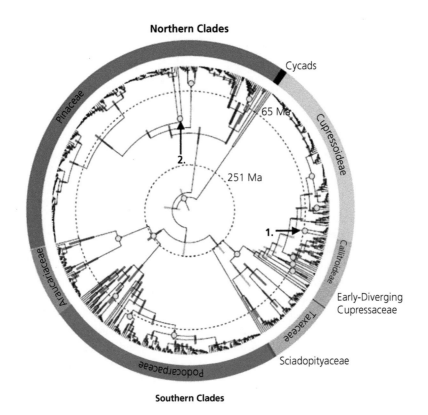

Fig. 7.3 Dated phylogeny for 489 presently existing conifer species. Cycads are included for comparative purposes in the construction methodology. Dashed circles with dates in Ma (million years ago) indicate boundaries between Paleozoic–Mesozoic and Mesozoic–Cenozoic. Yellow circles indicate divergent nodes in the classification used in calibration with fossils and described in the online Supplement to the article; arrows and numbers indicate nodes discussed in this text. Reproduced from Leslie A. B. et al. (2012).

Cretaceous–Paleogene (K–Pg) extinction (Willis and McElwain 2014, Chapter 8).

Leslie et al. report the median node ages for divergences in the Neogene is 5.2 My for northern hemisphere clades and 8.7 My for southern hemisphere clades. Species turnover rates in the northern hemisphere were estimated as generally greater than those in the southern hemisphere, and recently diverged northern hemisphere species are concentrated 'in regions of high conifer diversity such as mountainous areas of western North America and southern China'. They suggest that falling temperatures during the Cenozoic resulted in a change in mid to high latitude from warm to colder, drier, and more strongly seasonal climates which contributed to species changes. In the southern hemisphere regions of high conifer diversity have drifted

apart and northwards, leading to the persistence of relatively warm or wet habitats often moderated by oceanic climates and longer divergence times.

Chomicki et al. (2017) used the phylogeny developed by Leslie et al. to identify lineages and species with orthotropic (green lines in Fig. 7.4a) or plagiotropic (yellow) branching and rhythmic (purple lines in Fig. 7.4b) or diffuse (orange) growth. Rauh's model is identified as the ancestral architecture but with frequent transitions to plagiotropic branching (Fig. 7.4a) and diffuse branching (Fig. 7.4b). Chomicki et al. found 17 transitions from Rauh to Massart models with 4 reversions and 12 from Rauh to Attim with 1 reversion (Fig. 7.5). They report no simultaneous transitions to both plagiotropic and diffuse branching and suggest this may be due to either genetic constraints or that such

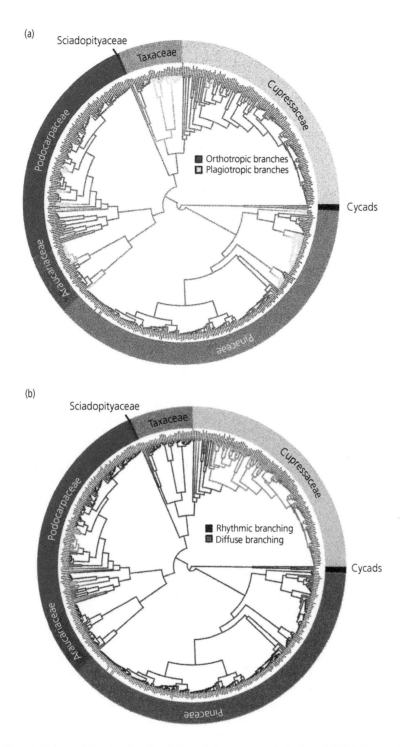

Fig. 7.4 Evolution of important characteristics of conifers. The phylogeny is that constructed by Leslie et al. (2012), but note that presentation of the sequence of lineages differs from that in Fig. 7.3. (a) Species with orthotropic branching, green; plagiotropic branching, yellow. (b) Species with rhythmic branching, purple; diffuse branching, orange. Reproduced from Chomicki G. et al. (2017) with permission from Oxford University Press.

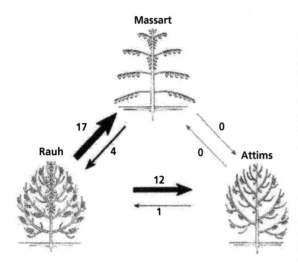

Fig. 7.5 Evolutionary progressions in conifer architecture with the number of transitions beside arrows as calculated from a calibrated phylogeny and orthotropic–plagiotropic and rhythmic–diffuse transitions (Fig. 7.4). Reproduced from Chomicki G. et al. (2017) with permission from Oxford University Press.

an architecture is not well adapted. Although evolution from orthotropic to plagiotropic branching occurred in each clade, evolution from rhythmic to diffuse growth did not occur in the Pinacæ or Araucariaceæ.

Shade-tolerant species are identified as belonging to both Massart's and Attim's models. In most coniferous forest of western North America, on the western slopes of the Cascade Mountains, *Pseudotsuga menziesii* (Rauh) is the long-lived pioneer (Ford and Ishii 2001) in association with shade-tolerant *Abies* and *Taxus* species (Massart), both *A. grandis* and *A. amabilis* and *T. brevifolia*, as well as *Thuja plicata* (Attim). The regularity in crowns of plants with Massart's model, where branches and foliage are planar, contrasts with the exploratory irregularity of Attim's; these models present contrasting ways of capturing light in shaded conditions. However, the generally numerical dominant shade-tolerant species in this forest type, *Tsuga heterophylla*, has the same form and is closely related to *Tsuga canadensis* suggested by Hallé et al. as following Mangenot's model, with mixed axes, plagiotropic and orthotropic, being produced by the same meristem.

Changes in environmental conditions seem likely to have led to evolution of particular architectures. Chomicki et al. suggest diffuse branching may have occurred earlier in evolution and they also found, through analyses of *Aloe* and related genera and arborescent monocaulus plants, that some architectures have evolved repeatedly and in some instances there have been many bidirectional transitions between models.

7.1.3 Foliage structure and display

Evolutionary changes occur in foliage as well as in branch architecture. Calibration points (yellow circles in Fig. 7.3) are based on first appearances in the fossil record of plants with characteristics of the new line of a lineage or genus. Point 1 (Fig. 7.3) marks the divergence of *Thuja* from *Thujopsis* dated as between 58 and 78 My by the fossil *Thuja polaris*. Both species have orthotropic branching and diffuse growth (Attim). Chomicki et al. note *Thuja* has leaflets that are broad and flattened in regular splays, unlike closely related *Thujopsis* which has similar but irregular splays. Cone differences are also present.

Point 2 (Fig. 7.3), the divergence between *Picea* and *Cathaya*, was based on the fossil *Picea burtonii* from the early Cretaceous (140–133 My) using morphological and anatomical characters of the seed cone. Both *Cathaya* and *Picea* have orthotropic branches and rhythmic growth, but there are differences in shoot and foliage structure between *Cathaya argyrophylla*, the single surviving species of the genus, and *Picea*.

Within the genus *Picea* there are substantial differences in both foliage structure and foliage display, which include both how foliage elements, i.e. needles, are arranged on shoots and how shoots are arranged on branches; examples are shown in Fig. 7.6. Some of these features correlate with environmental conditions of species habitats.

The genus *Picea* is considered to have originated in North America and Shao et al. (2019) propose there have been four dispersals between North America and Eurasia and dispersal to Asian islands. Using an extension of molecular dating methods Shao et al. define eight clades within the genus, with *P. breweriana*, which has a current geographic distribution restricted to western mountains on the California–Oregon border, as the single species forming the basal clade. The other seven clades form a circumpolar North American–Eurasian distribution.

Needle shape is used in morphology-based identification keys for *Picea*, notably whether they are flattened or quadrangular in cross section and, in addition, the amount and position of stomata on the needle (Dallimore and Jackson 1961) (Fig. 7.6). Wang et al. (2017) conducted a multivariate climatic analysis, using eight variables, of the distribution of morphological characters in the genus. They found the first divergence was between flattened needles, associated with wetter conditions, a narrow temperature annual range, and somewhat higher elevation, and quadrangular needles, associated with drier conditions, a wider temperature annual range, and lower elevation. Shao et al. conclude that flattened needles, along with closely arranged seed scales, 'are ancestral states, whereas quadrangular needles and closely arranged seed scales have evolved many times'. Species with flattened needles can also evolve in a clade that has a basal member with quadrangular needles.

A wide range of variation is also seen in size and pattern of distribution of needles along shoots (examples in Fig. 7.6), but this has not been analysed in relation to environmental variation. Branch posture in *Picea* can be pendulous, often in species with mountain habitats, e.g. *P. breweriana* and *P. smithiana*, and so is possibly related to avoidance of snow breakage.

7.2 Causes and effects of variation within an architectural model

At evolutionary time scales the development of an architecture is a dynamic process. The feedstock of evolutionary change is variation within a species and variation in realized architecture can occur between individuals of the same species. One cause can be the result of genetic differences in organogenesis, extension and foliage characters that affect development and growth. Understanding of how architecture affects growth requires taking into account its relationship with foliage amount and distribution.

Clonal differences in crown structure in young, open-grown trees of *Picea sitchensis* and *Pinus contorta* were potentially of interest for selection and breeding for plantation crops (Cannell et al. 1983). In both species young plants of sparsely branched clones produced more stem wood and between 1.5 and 2.0 times as much stem wood per unit of foliage as heavily branched clones. Differences in crown structure (Fig. 7.7) could possibly be related to alternative strategies to meet competitive environments, with bushiness being a 'fight' type and tallness a 'flight' type. Analysis of crown structures of *P. sitchensis* clones in two contrasting environments illustrates relationships between component processes that produce these differences in foliage amount and distribution and variation within the same basic architectural model. Their differences have effects on growth.

7.2.1 The architectural model of *Picea sitchensis*

Picea sitchensis follows Rauh's model, with three categories of axis: the leading shoot, whorl branches, and interwhorl branches (Fig. 7.8; see also Fig 2.2). The growth module is the foliated shoot that extends in a single year, with the basic structure of this module the same for each axis category. An apical bud has the potential to extend growth in the same direction as its parent, while a cluster of buds at the base of the apical bud form a whorl, and buds distributed along the length of the module are interwhorl buds. The realization of this potential structure depends upon the level of growth activity of the module.

Modules are orthotropic, but as whorl branch growth continues they typically become increasingly declined from the vertical (Cochrane and Ford 1978). Whorl laterals are radially spatially even; interwhorl laterals occur in groups that are spatially evenly distributed along the module. Typically. the numbers of both whorl and interwhorl laterals decline with increasing order and/or depth in the canopy of the parent module.

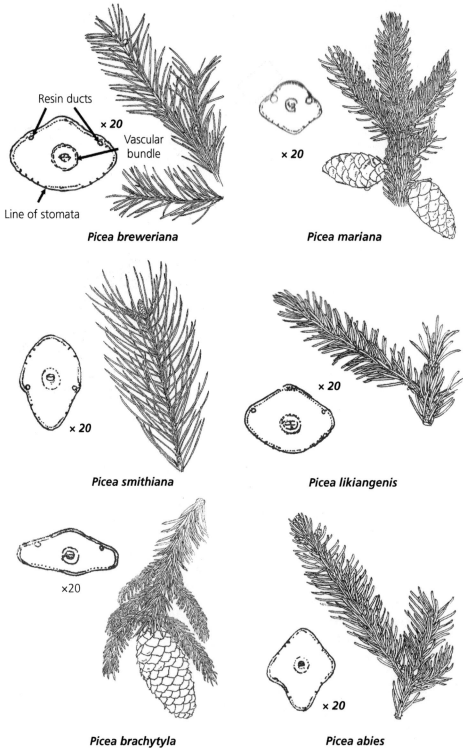

Resin ducts

Vascular bundle

Line of stomata

× 20

Picea breweriana

× 20

Picea mariana

× 20

Picea smithiana

× 20

Picea likiangenis

×20

Picea brachytyla

× 20

Picea abies

7.2.2 Organogenesis in above ground *Picea sitchensis*

Organogenesis and bud development, followed by bud growth into a foliated shoot, take place over three years (Fig. 7.9). In Year 1 a dome of tissue forms at the apex of the overwintering bud that will extend in Year 2 (Cannell and Willett 1975). In Year 2 the overwintering bud extends and forms the parent module and the dome of tissue becomes a new apical overwintering bud (and itself has an apical dome of tissue that will extend in Year 4). The apical overwintering bud contains foliage initials which develop into needles as the shoot extends in Year 3 and during this extension terminal and interwhorl buds develop along the module.

7.2.3 Variation in architecture and growth within young *Picea sitchensis*

Sheppard and Ford (1986) investigated plant structure and its relationship with growth for replicates of four clones of Sitka spruce on two sites of likely contrasting productivity: an old agricultural field (altitude 184 m; latitude 55° 55′ N) and a north-facing upland forest site (310 m; 55° 40′). Trees from different origins in the natural range of the species, referred to as provenances, have been widely studied and are considered as having different genetics, particularly, as in this case, where they come from quite widely different latitudes (Cahalan 1981; Xu et al. 2000).

Two clones were of Queen Charlotte Island origin (54° N), in Fig. 7.7 called QCI and 8010, the latter being a candidate elite or 'plus' tree (Cornelius 1994)

for a United Kingdom Forestry Commission breeding programme. These two were selected because a previous study on the agricultural field (Cannell et al. 1983) showed them tall with short branches and to have percentages of stem wood by total weight, which is of commercial interest, of 43 and 48% respectively. The other two clones, called here Cordova (origin 61° N) and Sitka (57° N), had stem wood percentages on the agricultural field of 39 and 26% respectively and both were shorter than the two more southerly clones but Sitka had similar total biomass to them while Cordova had less.

Neither site was fertilized prior to this trial. However, the soil on the agricultural field was a loam (pH 5.2) with higher concentrations of available N and P than the forest soil (pH 4.6) (Sheppard and Cannell 1985). Planting on both sites was at 1-m spacing in rows 1.5 m apart and measurements were made when trees on the two sites appeared similar in overall structure on seven-year-old trees on the agriculture site and nine-year-old trees on the forest site. Productivity, in terms of grams of dry weight produced per gram of dry weight of foliage, was estimated for the final year of growth before harvesting (Table 7.1) (Sheppard and Ford 1986).

The four clones had different relationships between the components of branch structure and in the foliage and wood weights on branches (Table 7.1), suggesting both genetic and environmental effects. The two clones of more southern origin were tallest on the agricultural field—QCI (5.2 m) and 8010 (5.9 m) compared with Cordova (4.3 m) and Sitka (3.1 m)—although clones were of similar height at the forest site— respectively 3.3, 3.1, 3.2, and 2.2 m. Mean total

---

Fig.7.6 *(Continued)* Structure of foliated shoots and leaf (needle) cross-sectional shape of selected *Picea* species. Needle cross sections at 20 times actual size with a central vascular tissue. *Picea breweriana*: needles flattened, about 2.5–3 cm long, with lines of stomata on lower surface on each side of midrib; Siskiyou Mountains, California–Oregon border. *Picea mariana*: needles quadrangular in section, about 1.25 cm long, and crowded on shoot, with one to four lines of stomata on each side; northern North America. *Picea smithiana*: needles about 3.75 cm long, distributed all round the shoot, incurved and pointing forward, quadrangular in section, with two lines of stomata on all four sides; western Himalaya. *Picea likiangenis*: needles on upper side of shoot pointing forward and overlapping, needles on lower side spreading in two opposite ranks, up to 2 cm long, with few stomata lines; west Szechuan mountains. *Picea brachytyla*: needles crowded and closely overlapping on upper side of shoot, those on lower side in two opposite ranks, 1.25–2 cm long, flattened but slightly ridged on surface, with two broad bands of stomata on lower surface; mountains of Hubei and Szechuan provinces. *Picea abies*: needles on upper side of shoot overlapping and pointing forward, those on lower side spreading right and left, 1.25–2.5 cm long, rhombic in cross section, with two or three lines of stomata on each side of needle; Northern and Central Europe distribution. Reproduced from Dollimore W. and Jackson A. B. (1961) with permission from Taylor and Francis, London.

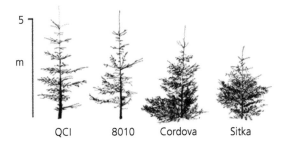

Fig. 7.7 Outline of individual trees of *P. sitchensis* drawn from photographs of the clones growing in an agricultural field.

above-ground tree weight for all clones taken together was 15% greater on the agricultural field than the forest site, but this was due to greater growth by QCI and 8010 of 53 and 82% respectively, while Cordova actually had 27% *less* weight on the agricultural field. Total above-ground production by QCI and 8010 over the year preceding harvest on the agricultural field was 1.31 and 1.41 greater than on the forest site: corresponding values for Cordova and Sitka were 0.66, a *reduction*, and 1.07 respectively. Overall, stem production, in terms of both weight and proportion, was greater on the agricultural field, although this was related to the 12–14% more total production for QCI and 8010

which had greater stem wood proportions than Cordova and Sitka which both had similar stem wood amounts at both sites.

There was a general inverse relationship between foliage amount and production per unit of foliage (Table 7.1). At the forest site 8010 had the least amount of foliage and greatest above ground productivity per unit of foliage. On the agricultural field this clone had 2.75 times as much foliage as it had on the forest site and greater growth, but it had the lowest production per unit of foliage of all clones on the agricultural field. The greater total dry weight increment on the agricultural field by QCI and 8010 reflects greater foliage amount than on the forest site by factors of 1.71 and 2.74 respectively, which offset reduced production per unit of foliage by factors of 0.77 and 0.51 respectively. Sitka and QCI also had similar inverse relationships between foliage and productivity per unit of foliage; however, Cordova had greater total foliage *and* production per unit of foliage at the forest site.

A decrease in dry weight increment per unit of foliage with increasing weight of foliage on a young tree might be due to an increase in self-shading. The two clones with greatest rate of production per unit of foliage at the forest site, QCI and 8010, had

Table 7.1 Last annual biomass increments of stem wood, branch wood, and foliage and percentage of above ground increment for each of four clones of *P. sitchensis* at a forest site and on an agricultural field. Foliage weight estimated at the start of growth and productivity of that foliage shows considerable differences between both clones and sites. Calculated from Sheppard and Ford (1986).

	Last annual increment (g tree^{-1})						Estimated foliage weight prior to growth (g tree^{-1})	Total increment per estimated foliage (g g^{-1}y^{-1})
	Stem wood	%	Branch wood	%	Foliage	%	Total		
Forest site									
QCI	601	32	580	31	696	37	1877	718	2.16
8010	579	30	842	44	478	26	1899	651	2.96
Cordova	711	20	1517	43	1306	37	3535	1741	2.03
Sitka	507	17	1411	47	1057	36	2974	2008	1.48
Agricultural field									
QCI	904	37	739	30	821	33	2464	1225	2.01
8010	1185	45	665	25	804	30	2654	1787	1.49
Cordova	708	30	750	32	870	37	2328	1316	1.77
Sitka	611	19	1143	36	1416	45	3170	1552	2.04

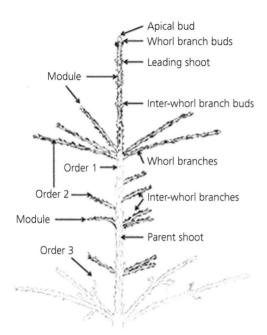

Fig. 7.8 Components of the branching structure of *P. sitchensis*. The leading shoot of the tree has apical whorl and interwhorl buds that all grow into shoots in the same year. A parent leading shoot, drawn with a lighter line in this figure, supports a new leading shoot and whorl and interwhorl branch shoots. Each new shoot, of whatever type, is a module that may itself have buds of the apical whorl and interwhorl type that potentially could grow. Branch order refers to the sequence of successive branching axes, irrespective of age, with order 1 being the vertical axis, order 2 branches that come from that axis, etc. Generally the numbers of buds set, and their actual growth, decrease with increase in the order of the branch and with increasing depth in the crown. Reproduced from Cochrane and Ford (1978) with permission from Wiley.

sparser crowns than Cordova and Sitka and had foliage held closer to the trunk: mean total lengths of module per whorl branch were 4.38, 5.35, 10.28, and 9.12 m (QCI, 8010, Cordova, Sitka). On the agricultural field total length of modules per whorl branch was greater for 8010 (8.02 m), although little different for both QCI (4.10) and Sitka (9.5 m). For Cordova both foliage weight per unit length and total length of modules per whorl branch were *less* on the agricultural field (0.41 kg m^{-1} and 0.769 m), as was above-ground production. Generally, QCI and 8010 had greater foliage retention and nutrient use efficiency in terms of stem wood produced per nutrient uptake (Sheppard and Cannell 1985). So, although all four clones have the same basic module and architectural model as given in

Fig. 7.8, quantitative differences in organogenesis and extension exist between individual clones that result in differences in growth rate and development of plant form. Of particular interest is the lack of increase on the agricultural site for Cordova while other clones increased.

7.2.4 The dynamics of architecture and growth in young *Picea sitchensis*

Dynamics of the growth of *P. sitchensis* can be summarized as interactions between processes that may increase growth rate and processes that may decrease it. Three processes, boxed numbers [1], [2], and [3] in Fig. 7.10, can operate to increase growth as the young tree increases in size:

[1] Increase in total length of shoots depends upon the numbers of buds that form, their duration and rate of growth, and the supply of resources required to produce the growth. Shoot length provides the scaffold structure for foliage and future bud development. Numbers of buds produced and the duration of shoot growth are both under genetic and environmental influences.

[2] Increase in foliage amount is determined by shoot length increment and needle number and their sizes and which require resources for growth.

[3] As foliage amount increases there is increase in production of resources available for growth, but this increase depends upon production per unit of foliage. The resources considered in Fig. 7.10 are restricted to products of photosynthesis.

Generally, three components may operate to reduce the rate of increase in growth:

[4] The rate of production per unit of foliage may decrease as foliage amount increases; an important contributing reason may be increased shading of older foliage as the crown increases in size.

[5] Resources are required for wood in the growth of new shoots and in diameter growth of support structures: branches, trunk, and roots.

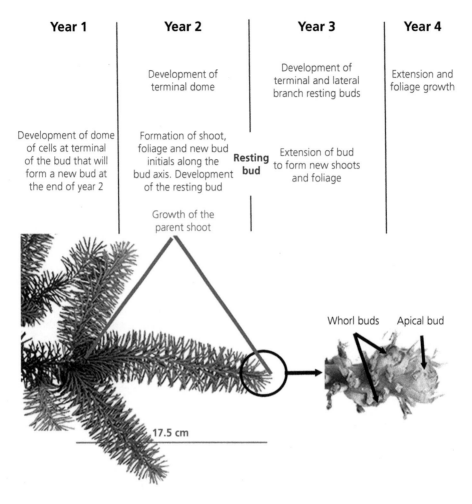

Fig. 7.9 The sequence of organogenesis and extension of foliated module production in *P. sitchensis* trees. Bud development (blue text) starts (Year 1) in the year before its parent shoot extends with growth at the distal end of an overwintering bud that will extend in Year 2. During Year 2, as extension of the parent shoot is completed, which is the stage illustrated in the lower part of the figure, this dome of tissue differentiates to form the bud that overwinters between Years 2 and 3 and contains needle initials and young stem tissue. An apical and two whorl buds at the distal end of the parent shoot are shown with bud scales (brown tissue) starting to develop. As these buds extend in Year 3 a new apical and lateral buds are formed for shoots that will extend in Year 4 (red text).

Generally, most biomass increment of cambial growth over the complete branching structure occurs after shoot growth, but cambial growth of the extending shoot and fine root growth increment may be a requirement during extension.

[6] Foliage senescence and death reduces foliage amount.

Genetics obviously determines growth processes in a number of ways and the interactions outlined in Fig. 7.10 can produce apparently similar features of growth in different ways. Although QCI and 8010 had large and similar stem weight proportions at both sites, these two clones had different crown structures despite some visual similarity. The dominant features for QCI were the small numbers of modules and a small total module length supporting a large weight of foliage (per unit branch length) together with a small rate of foliage loss. These modules were comparatively heavy in terms of wood weight per unit length and can be assumed

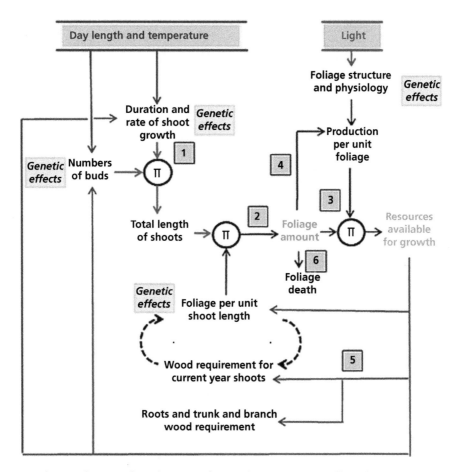

Fig. 7.10 Interactions between above ground growth processes of young Sitka spruce, on an annual basis, that may increase growth rate through an increase in foliage amount, and so available resources for growth, or may decrease it. Boxed numbers refer to component processes defined in the text. Π indicates a multiplicative interaction. Some genetic and environmental effects on the process are discussed in the text.

to have a large diameter of wood, although their short length meant that total dry matter used in branch wood production was low. 8010 had more whorl branches at both sites than QCI (mean proportional difference of 1.8), supporting about 2.5 times as many modules. However, the total branch wood weight differences between these clones were only 1.4 (8010/QCI) for whorl and 1.7 for interwhorl branches and wood per unit branch length was less for 8010 than QCI. Clone 8010 only carried 30% of the weight of foliage per unit module length of QCI. At the forest site 8010 incremented proportionally more dry matter to the production of branch wood than to foliage, although on the agricultural field the proportions and amounts of

branch wood and foliage were similar. Comparing the agricultural field with the forest site, both QCI and 8010 had greater module lengths by factors of 1.3 and 1.9 whorl, 1.3 and 3.0 interwhorl. Total numbers of modules were not substantially greater for QCI but were for 8010: 1.1, 1.4 (QCI, 8010) whorl; 1.0, 2.1 interwhorl.

Productivity does not necessarily increase with what might normally be considered as better growing conditions but can depend upon how those conditions affect the architecture of the plant. The finding of greater growth on the agricultural field compared to the forest site for three of the four clones is what might reasonably be expected considering likely greater nutrient availability of the agricultural soil and warmer conditions at the lower altitude—but the growth of Cordova provides a contrast. Although its stem wood weights were identical at the two sites, total above ground dry weight was considerably *less* on the agricultural field compared to the forest site. The lesser foliage amount at the agricultural field relative to the forest site for Cordova (0.72) was due to a reduction in number of whorl branches (by a factor of 0.54) even though the number of interwhorl branches was greater (1.64). Cordova had the smallest whorl branch length at both sites and length increased less between the forest site and agricultural field (1.18) than the other clones (mean 1.26). The efficiency of wood production per unit of foliage for Cordova was greater on the agricultural field where there was less foliage.

An explanation for the reduced number of whorl branches but greater number of interwhorl branches for Cordova on the agricultural field relative to the forest site can be found in the distribution of resources between leader extension and branch bud formation and the phasic development of leaders. The initials that will produce lateral buds from the leading shoot form while the leading bud develops, i.e. in the year before its extension as a leading shoot and so two years before extension of the lateral shoots themselves (Cannell and Bowler 1978) (Fig. 7.9). Proximal buds develop before more distal ones and Cannell and Bowler found the initials for all buds, except those that would produce the terminal whorl of branches, had formed before shoot extension started but while the bud was swelling. The terminal whorl of buds developed on the newly forming apical bud tissue while the shoot supporting it was extending.

The time of bud formation in Sitka spruce is closely linked with latitude of seed origin and shows a north–south progression. Malcolm and Pymar (1975) found through experiments in controlled environments that the period of shoot extension could be extended by increased temperature for southerly but not more northerly provenances. It is possible that the more favourable conditions for growth on the agricultural site led to greater bud production and hence more branches. However, for Cordova, a northerly provenance, this bud formation process was truncated by the shorter day length than occurs during the shoot extension period at its native latitude, so that while interwhorl bud formation along the more proximal part of the shoot increased in response to better conditions, development of a terminal whorl of buds was restricted.

In these clonal trials with young trees, increased foliage amount appears to have been instrumental in achieving higher growth, but a noticeable feature is the inverse relationship between foliage amount and the efficiency of foliage in producing increment in terms of gram of dry weight per unit weight of foliage. This is likely related to self-shading within the plant (e.g. Anten and Hirose 1998), although no calculations of this effect have been made for Sitka spruce. At this stage, when plants are open-grown and have not come into intense competition, achieving large foliage amount outweighs foliage efficiency in producing the growth increment, but it seems reasonable to expect that there are limits to this.

7.3 Conclusion

Existence of convergent evolution of plant architecture as defined for models proposed by Hallé et al. suggests they encapsulate important features affecting growth and/or survival. Patterns of repeated evolution and reversions in some architectures and occurrence of some architectural models in herbaceous plants as well as trees suggest a potentially wide application for the theory advanced by Hallé et al.

An architectural model may define some basic components of how a species functions and grows in a particular environment. For instance, Rauh's model defines a plant with upward growing branches, typical of plants that colonize open ground or patches and grow to form closed canopies—pines are notable examples. Attim's and Massart's models apply to species that are shade tolerant but exist in the competitive environment in different ways: in closed canopies *Abies* species (Massart) produce a dense crown that is difficult for other species to penetrate (Kennedy 2010); *Thuja* (Attim) produces branches that appear to explore for light.

The Hallé et al. models do have limited explanatory power. Some differences between species with the same model may be due to architectural differences not specified by the models of Hallé et al. For example, within species of *Zingiberales*, all classified as following Tomlinson's model with rhizomatous growth, there are substantial differences in architecture that contribute to differences in growth and ecology (Chomicki 2013) and these can be defined by the type of branching of the rhizome.

Three components of plant architecture that may affect growth are not included in the Hallé et al. system:

1. Foliage structure and display: the architecture of a plant is the product of its growth and it influences how future growth will be made—but understanding the dynamics of this relationship requires that more than just branching morphology be considered. The analysis of *Picea* clones shows that foliage and its display should be considered in a functional definition of architecture. Morphological development of a plant, which is the foundation of its architecture, is affected by many component processes and this can result in considerable differences in growth, as illustrated in clones of *Picea*. Defining morphological plasticity as an integral component of architectural models (Chapter 8) rather than considering it as an exogenous effect on an endogenous model as suggested by Hallé et al. also requires extending the definition of architecture to foliage since foliage as well as branching structure can show plasticity.

2. Architecture needs to be integrated with plant development to understand growth: growth is the result of balance between positive and negative processes that determine its rate. Controls expressed at the level of the module (Fig. 7.9) determine build-up of foliage and so contribute to future growth. In *P. sitchensis* both initiation and cessation of growth are triggered by environmental conditions but with trigger points determined by genetics and physiological conditions—the potential for growth is under genetic control. Details of structure in the architecture may affect resource acquisition and module increment operates under a control of resource availability which in turn can be affected by environmental conditions and genetic interactions affecting plasticity. Mathieu et al. (2008, 2012) illustrate how oscillations in growth, of the type observed in fruit or branch appearance, can be generated. Organ growth can reduce resource availability below the threshold for new organ formation so there is a growth pause until resources are replenished.

3. Variation in architecture can occur between individuals within a species. This is seen in both the amount of foliage supported by different clones of *P. sitchensis* and crown structure which affects its display and receipt of light.

These additional considerations are required for architectural models to explain some essential features of plant growth and provide understanding of the extent to which developmental stability of plant architecture may be achieved.

7.4 References

Anten, N. P. R. and Hirose, T. 1998. Biomass allocation and light partitioning among dominant and subordinate individuals in *Xanthium canadense* stands. *Annals of Botany*, 82, 665–673.

Cahalan, C. M. 1981. Provenance and clonal variation in growth, branching and phenology in *Picea sitchensis* and *Pinus contorta*. *Silvae Genetica*, 30, 40–46.

Cannell, M. G. R. and Bowler, K. C. 1978. Spatial arrangement of lateral buds at the time that they form on leaders of *Picea* and *Larix*. *Canadian Journal of Forest Research*, 8, 129–137.

Cannell, M. G. R., Sheppard, L. J., Ford, E. D. and Wilson, R. H. F. 1983. Clonal differences in dry-matter distribution, wood specific-gravity and foliage efficiency in *Picea sitchensis* and *Pinus contorta*. *Silvae Genetica*, 32, 195–202.

Cannell, M. G. R. and Willett, S. C. 1975. Times and rates at which needles are initiated in buds on differing provenances of *Pinus contorta* and *Picea sitchensis* in Scotland. *Canadian Journal of Forest Research*, 5, 367–380.

Cannell, M. G. R., Sheppard, L. J., Ford, E. D. and Wilson, R. H. F. 1983. Clonal differences in dry-matter distribution, wood specific-gravity and foliage efficiency in *Picea sitchensis* and *Pinus contorta*. *Silvae Genetica*, 32, 195–202.

Chomicki, G. 2013. Analysis of rhizome morphology of the Zingiberales in Payamino (Ecuador) reveals convergent evolution of two distinct architectural strategies. *Acta Botanica Gallica*, 160, 237–252.

Chomicki, G., Coiro, M. and Renner, S. S. 2017. Evolution and ecology of plant architecture: integrating insights from the fossil record, extant morphology, developmental genetics and phylogenies. *Annals of Botany*, 120, 855–891.

Chomicki, G., Staedler, Y. M., Bidel, L. P. R., Jay-Allemand, C., Schonenberger, J. and Renner, S. S. 2018. Deciphering the complex architecture of an herb using micro-computed X-ray tomography, with an illustrated discussion on architectural diversity of herbs. *Botanical Journal of the Linnean Society*, 186, 145–157.

Cochrane, L. A. and Ford, E. D. 1978. Growth of a Sitka spruce plantation: analysis and stochastic description of the development of the branching structure. *Journal of Applied Ecology*, 15, 227–244.

Cornelius, J. 1994. The effectiveness of plus-tree selection for yield. *Forest Ecology and Management*, 67, 23–34.

Dallimore, W. and Jackson, A. B. 1961. *A Handbook of Coniferae Including Ginkoaceae*, Third Edition 1948, reprinted with corrections. London, Edward Arnold.

Ford, E. and Ishii, H. 2001. The method of synthesis in ecology. *Oikos*, 93, 153–160.

Hallé, F., Oldeman, R. A. A. and Tomlinson, P. B. 1978. Tropical Trees and Forests: An Architectural Analysis. New York, Springer.

Kennedy, M. C. 2010. Functional–structural models optimize the placement of foliage units for multiple whole-canopy functions. *Ecological Research*, 25, 723–732.

Leslie, A. B., Beaulieu, J. M., Rai, H. S., Crane, P. R., Donoghue, M. J. and Mathews, S. 2012. Hemisphere-scale differences in conifer evolutionary dynamics. *Proceedings of the National Academy of Sciences of the United States of America*, 109, 16217–16221.

Leverenz, J. W., Whitehead, D. and Stewart, G. H. 2000. Quantitative analyses of shade-shoot architecture of conifers native to New Zealand. *Trees: Structure and Function*, 15, 42–49.

Malcolm, D. C. and Pymar, C. F. 1975. The influence of temperature on the cessation of height growth of Sitka spruce (*Picea sitchensis* (Bong.) Carr.) provenances. *Silvae Genetica*, 24, 129–132.

Mathieu, A., Cournède, P.-H., Barthélémy, D. and De Reffye, P. 2008. Rhythms and alternating patterns in plants as emergent properties of a model of interaction between development and functioning. *Annals of Botany*, 101, 1233–1242.

Mathieu, A., Letort, V., Cournède, P.-H., Zhang, B. G., Heuret, P. and De Reffye, P. 2012. Oscillations in functional structural plant growth models. *Mathematical Modelling of Natural Phenomena*, 7, 47–66.

Sauquet, H. 2013. A practical guide to molecular dating. *Comptes Rendus Palevol*, 12, 355–367.

Shao, C. C., Shen, T. T., Jin, W. T., Mao, H. J., Ran, J. H. and Wang, X. Q. 2019. Phylotranscriptomics resolves interspecific relationships and indicates multiple historical out-of-North America dispersals through the Bering Land Bridge for the genus *Picea* (Pinaceae). *Molecular Phylogenetics and Evolution*, 141, 106610.

Sheppard, L. J. and Ford, E. D. 1986. Genetic and environmental control of crown development in *Picea sitchensis* and its relation to stem wood production. *Tree Physiology*, 1, 341–354.

Sheppard, L. J. and Cannell, M. G. R. 1985. Nutrient use efficiency of clones of *Picea sitchensis* and *Pinus contorta*. *Silva Genetica*, 34, 126–132.

Sheppard, L. J. and Ford, E. D. 1986. Genetic and environmental control of crown development in *Picea sitchensis* and its relation to stem wood production. *Tree Physiology*, 1, 341–354.

Véron, E., Vernoux, T. and Coudert, Y. 2021. Phyllotaxis from a single apical cell. *Trends in Plant Science*, 26, 124–131.

Wang, G. H., Li, H., Zhao, H. W. and Zhang, W. K. 2017. Detecting climatically driven phylogenetic and morphological divergence among spruce (*Picea*) species worldwide. *Biogeosciences*, 14, 2307–2319.

Willis, K. J. and McElwain, J. C. 2014. *The Evolution of Plants*, Second Edition. Oxford, Oxford University Press.

Xu, P., Ying, C. C. and El-Kassaby, Y. A. 2000. Multivariate analyses of causal correlation between growth and climate in Sitka spruce. *Silvae Genetica*, 49, 257–263.

CHAPTER 8

Growth within foliage canopies

8.1 Introduction

Foliage canopies develop and are maintained by plants sufficiently close together that their crowns become and remain contiguous. Canopy structure and the environment it produces changes continually as individual plants grow, both vertically and laterally, and come to shade their neighbours and be shaded by them. The dynamics of this process are determined by the physiology and architecture of plants and how these respond to change in canopy environment and within plants themselves. This process has been studied from different perspectives.

One perspective has long had the objective of understanding the basis of stand production or crop yield and considers the canopy as a unit, typically expressed in terms of leaf area index (L), with stand production related to its foliage amount and interception or absorption of radiation. For example, Black (1964) defines factors affecting the annual cycle of L and production for a crop; Sinclair and Muchow (1999) review the approach of calculating production in terms of radiation use efficiency. Individual plants are not frequently considered in these approaches, although they are in practical crop husbandry (e.g. for cotton; Bednarz et al. 2006).

A second perspective has the objective of analysing the process of competition, where some plants succeed while others do not and may ultimately die, termed self-thinning. Weller (1987) reviews the relationship between the rates of plant mortality and mean plant weight of survivors in single species stands undergoing competition and shows that this relationship differs between plant types.

A third approach has focussed on the effects that change in the spectral quality of light within a canopy can have on morphology and development. With increase in depth in canopies change occurs in the red region of the spectrum: foliage absorbs red light (R; 660 nm) preferentially to far red (FR; 730 nm) and decrease in R:FR ratio affects the phytohormone phytochrome. This can produce a relative increase in height growth for shaded plants that typically grow in sunny environments: Morgan and Smith (1978) for *Chenopodium album* and *Sinapsis alba*; Ballaré et al. (1994) for tobacco. This type of response has been called a 'shade avoidance' phenomenon (Smith and Whitelam 1997; Franklin 2008). However, the effect of change in spectral distribution is not restricted to height growth but can stimulate plasticity in other characters affecting growth.

Contributions of these and related approaches to understanding of growth in canopies are reviewed and illustrate how variation between plant types and/or environmental conditions can determine how control processes function in different canopies. This is discussed in relation to four aspects of canopy growth:

1. When competition is sufficiently intense, repeated contests between larger plants result in a spatial equalization of large individuals. This equalization process is fundamental to determining the plant's community and canopy structure and can be affected by plant architecture.
2. Morphological plasticity of different types and for different characters occurs on plants growing in canopies. These vary between plants with different architectures and foliage.

The Dynamics of Plant Growth. E. David Ford, Oxford University Press. © E. David Ford (2023). DOI: 10.1093/oso/9780192867179.003.0008

3. As a canopy grows, different processes can become paramount in response to the contests of competition and so affect growth.

4. Some mitigation of competition effects can occur through plasticity of plant architecture and determine the growth of the stand as a whole. This is analysed through the effects of foliage structure and plasticity on light interception and yield in different maize hybrids.

8.2 Community and canopy structure

Much work on plant competition has been conducted in even-aged, single species stands whether produced by natural regeneration or through agriculture. When individuals are closely spaced then, typically, after early initial growth the stand will go through a self-thinning phase (Weller 1987). This can be studied by measuring relative growth rates (RGR) and mortality rates of individuals of different sizes; generally, larger plants have greater RGR while smaller plants have greater mortality (Ford 1975).

8.2.1 Spatial interactions

The change most indicative of competition is development of spatial evenness of survivors or of dominant plants if the stand has not progressed to self-thinning. Jack pine (*Pinus banksiana*) regenerates through release of seeds from serotinous cones that open immediately following a crown fire producing even-aged stands. In a 65-year-old pure stand Kenkel (1988) mapped locations of both live and dead trees which tend not to rot in this environment and so can be censused. Live + dead trees ($n = 1,375$), which taken together represent the stand establishment structure, were randomly distributed, but live trees alone ($n = 459$), that had survived the competition process up to that time, were significantly regularly distributed with greatest deviation from random over the distance of 0.5–3.5 m. Kenkel estimated that for a radius of 3.5 m each tree was competing with an average of six neighbours.

Spatial evenness indicates a degree of environmental control over growth. Plants become relatively large in a stand because they have advantageous outcomes over the sum total of competitive interactions with their neighbours. These outcomes depend upon the balance between capacity to exert competitive influence and growth in response to resources obtained. A plant surrounded by generally large plants is likely not to grow as much as one surrounded by lesser sized plants. This is a spatial equalization process.

8.2.2 Crown asymmetry

Competition for light occurs as a series of contests in three dimensions where plant crowns interact for access to light above and to the side of them. Growth of crowns of the annual *Xanthium canadense* (now *X. strumarium*), the common cocklebur, was preferentially towards the direction of light (Umeki 1995). Individuals had been transplanted with random spatial arrangement onto a 5×10 m plot at a density of 0.5 individuals m^{-2} and had a final mean height of 1.94 m with no mortality in the stand. A crown vector that joined the stem base position with the centroid of its crown position had a mean value of 43 cm after 114 days, i.e. crown centres were substantially displaced from the randomly distributed planting positions and as this population grew the spatial pattern of crown centres became regular.

Crown asymmetry has been found in a number of tree species (Ford 2014). Typically the asymmetry is related to neighbours and in some cases strongest neighbours as defined by size and distance (Brisson 2001). In naturally regenerated *Pinus sylvestris* the displacement of tree crowns relative to their trunks increased as stands aged. Uria-Diez and Pommerening (2017) mapped the position of trees and their crown projections for a series of Scots pine stands at different developmental stages (Fig. 8.1a). For the stand with the tallest trees (mean 29.84 m) and largest diameter at breast height (dbh) (46.33 cm) the mean crown displacement was 0.92 m but values ranged up to greater than 3 m (Fig. 8.1c), while for the stand with shortest trees (13.64 m) and smallest dbh (20.64 cm) mean crown displacement was 0.35 m but ranged up to just less than 3 m. Uria-Diez and Pommerening conclude that crowns tended to fill towards empty space and that mature trees show most crown displacement and that wind and aspect did not have big effects in this forest, although wind can affect crown structure in some situations.

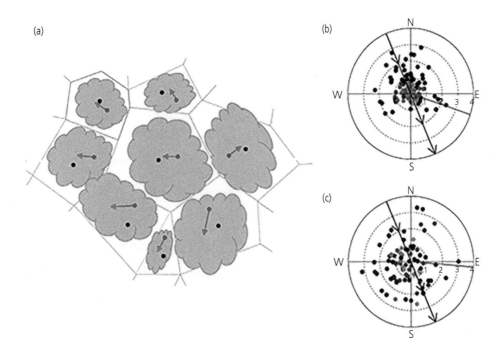

Fig. 8.1 (a) Schematic representation of tree crowns of *Pinus sylvestris* showing trunk location (blue), crown centres (end point of red arrows), and apparent space available for the individual defined by Voronoi polygons, grey lines, with tessellation centroids (black). (b) Circular plots of distance (m) and angle of displacement of crown centres from trunks for a young and (c) mature stand. Black arrows indicate mean wind direction, red line the plot aspect. Larger values of dbh in black, smaller in grey. Reproduced from Uria-Diez and Pommerening (2017) with permission from Elsevier.

8.2.3 Development of tiered canopies

Competition in canopies produces a distinctive vertical structure which is related to the development of horizontal patterns of plants and crowns. Bimodal distributions with distinct groups of large and small size plants have frequently been found (Turley and Ford 2011, Table 1). Turley and Ford demonstrate that these subpopulations can be reliably defined by fitting a bivariate (plant height and plant weight) mixture distribution using evolutionary computation. In multiple data sets, for different species and plants grown under different initial spacing, they found large- and small-plant subpopulations, respectively forming an upper and lower canopy.

Turley and Ford analysed data sets from time series and spacing experiments of *Tagetes patula* (Fig. 8.2a). In an experiment with plants from an initial spacing of 2 cm, after 28 days the large-sized subpopulation contained 35% of the originally planted individuals, 15% at 42 days, and 12% at 56 days. In each case horizontal separation of large plants was statistically significant.

The occurrence of a distinct upper canopy composed of spatially evenly distributed plants can be maintained, although with decreasing numbers of plants contributing to this *large* subpopulation. In *Tagetes patula* leaves and axillary shoots grow upwards and away from the main stem axis, and competition between *large* plants occurs in the volume of space above the current canopy. Between days 42 and 56 of a 2-cm initially spaced plant population, modal values of height for *large* increased from 9.05 to 11.9 cm, with a loss of 41% of individuals from the *large* subpopulation over the 14-day period.

Part of an actual 42-day lattice (Fig. 8.3) shows distribution of *large* and *small* plants and positions with dead plants. *Large* are generally evenly spaced as would be expected, but there are some *large*

(a)

(b)

Fig. 8.2 (a) An open grown *Tagetes patula* having a bush-like form, referred to in the text as having decurrent growth. Height 10.5 cm to main axis flower bud and 16.5 cm wide at the base of the plant. Leafy shoots, marked with arrows, are produced in the axils of leaves along the main stem. (b) An open grown *Chenopodium album*, referred to in the text as having excurrent growth, with leaves on the upper part of the shoot being small relative to those on the lower part. Height 8 cm from base to main shoot bud. Arrows mark axillary shoots.

plants with areas of greater overlap and overlap from more than one neighbouring *large* plant, e.g. plants A, B, C (Fig. 8.3). The number of encounters between *large* plants is likely to be greater where they are closer neighbours. The continued occurrence of spatially even distributions of *large* plants indicates that nearest neighbour distances increase over time. At 56 days, the total number of surviving plants is less than in *large* plants at 28 days, so there was competition between *large* plants over this time. Loss of *large* status would be expected by one plant from each of the pairs represented by A, B, and C. At 56 days surviving plants are spatially evenly distributed through the matrix of dead plants.

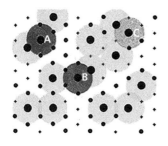

Fig. 8.3 The spatial arrangement of plants in an actual section of a 42-day triangular lattice planting of *T. patula*. *Large* plants are represented by large black circles, *small* plants by small black circles, and positions with dead plants by a small black dot. *Large* plants have become spatially even. A 2-cm-diameter coloured circle is drawn around each large plant. Only three of the large plants, those labelled A, B, and C, represent one of a couple of plants that are neighbours at a 2-cm distance. Reproduced from Ford (2014) "The dynamic relationship between plant architecture and competition". *Frontiers in Plant Science* 5: Article 275.

The prospect of a *large* plant continuing as *large* depends upon its proximity to all of its *large* neighbours (Fig. 8.3). Before there is substantial mortality, out-competed *small* plants are clustered round large plants and differences in size between *large*, which continue to grow, and *small*, which do, are sufficient to result in a bimodal distribution because *small* plants do not die immediately on being overtopped.

Initial density of plants affects the course of the competition process and resulting stand structures. After any time interval the number of dominants per area of ground is not constant across different initial plant densities. At 56 days, 2 cm initial spaced plants were taller than those at 4 cm, which may be attributed to greater density-induced height growth stimulation at 2 cm initial spacing (e.g. Ballaré et al. 1994). There were twice as many surviving plants per unit area at 2 cm initial spacing than at 4 cm, i.e. 95 (2 cm) to 52.5 (4 cm), although survivorship of initial plants was actually less—23.75% compared to 52.25%—and the respective areas of lateral influence of large plants were 24.5 and 71.1 cm^2 for 2 and 4 cm initial spacing, respectively.

The effects of canopy structure on competition can vary between species, which is not surprising given the variation that exists in branch and foliage structure. In the annual plant *Chenopodium*

album (lamb's quarters or fat hen) (Fig. 8.2b) Nagashima and Terashima (1995) found a two-tier canopy developed early in experimentally sown stands and did not change greatly over the life of the stand. Repeated measurements of individual plant heights in similarly established stands (Nagashima et al. 1995) showed that many plants ceased height growth early in stand development.

Nagashima (1999) calculated height rank of individual plants of *C. album* throughout the ~130 days of stand growth and estimated mark correlation between height at a sample time and final height. For all initial planting densities a plant's rank changed only in an initial short period: ~25 days after first emergence for initial density of 3,600 plants m^{-2} and ~43 days for 400 and 800 initial plants m^{-2} (Nagashima 1999, Fig. 1.) Nagashima concluded: 'the fate of a plant in a crowded stand is determined in the early stage of stand development'.

This conclusion differs from stands of both *T. patula* where *large* plants continuously declined in number and the development of stand structure in *P. banksiana* where Kenkel (1988) suggests the remaining live trees after 65 years (459 out of an initial 1,375) were competing with an average of six neighbours. An explanation for differences between *T. patula* and *C. album* may lie in a combination of their crown morphology and spacing. *Tagetes patula* has strongly decurrent growth (Fig. 8.3a) in which lateral growth of foliage is greater than terminal growth. *Large* plants at 42 days at 2 cm regular initial spacing produced leaves that extended horizontally >10 cm whereas median height growth was 2.4 cm over the 42- to 56-day interval. In contrast *C. album* develops markedly excurrent growth (Fig. 8.3b) as the plant grows. Leaves are alternate from the stem, and first leaves at the base of the plant are typically 3–7 cm long and 3–6 cm broad but towards the upper part of the plant are 1–5 cm long but only 0.4–2 cm broad. The plant develops as a spike. This suggests that the capacity for producing shade is greater for *T. patula* than *C. album*, although for *T. patula* this capacity may vary depending upon the spacing of the plants, e.g. being greater at 2 cm than 4 cm initial spacing.

8.3 Plasticity

As plants grow within canopies the shapes of their crowns change. This is the result of phenotypic plasticity (environmentally induced changes to the developing morphology). Two types of plasticity occur (Ford 2014): resource acquisition plasticity is the capacity of modules of foliage and branch growth to respond to a difference in available resources in a way that causes local differences in growth and so a change in shape; morphogenetic plasticity is the result of plant meristems and/or tissues responding to a changed spectral quality of light by changing or modifying the structure of growth.

8.3.1 Resource acquisition plasticity

Analysis of plastic responses of branch growth to variation in light quantity (Umeki and Seino 2003) is complicated by overall light conditions within a crown and potential competition for resources between branches.

Koike (1989) found shoots develop towards brighter light without there being a phototropic effect—the growth is due to utilization of the greater available resources. He surveyed seasonal growth from parental shoot tips in the upper crowns of two evergreen *Quercus* species, *Q. acuta* and the more shade-tolerant *Q. gliva*. Relative diffuse light intensity and its principal direction at each parent tip was estimated using hemispherical camera photographs and both shoot length increment and new shoots produced were measured. Analysis was made using a model (Fig. 8.4) that sought to relate growth amount and its direction to relative light amount and direction and taking gravity into account. For both species, shoot direction was significantly affected by geotropism but not correlated with the direction from which light was received. Extension growth was measured and the population of all measured shoots for each species was used to calculate magnitudes of vectors for growth (N in Fig. 8.4) in terms of the perpendicular component of light direction (L) and the perpendicular component of gravity in terms of negative geotropism (G). Koike concludes: 'Phototropism

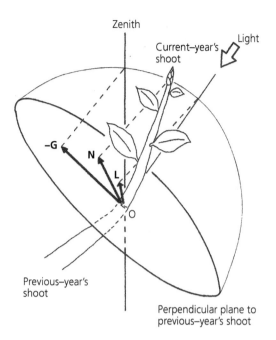

Fig. 8.4 Model for the analysis of shoot growth in relation to shoot orientation. 0 is the tip of a previous-year's parental shoot, L is the perpendicular component of the light direction, G is the perpendicular component of gravity (–G in this figure indices negative geotropism), and N is the perpendicular component of the current-year's shoot direction. Reproduced from Koike (1989) "Foliage-crown development and interaction in *Quercus gilva* and *Q. acuta*". *Journal of Ecology* 77(1):92–111, with permission from Wiley.

may not be very important in the upper canopies of forests.'

Koike found shoot production increased with increasing light received for both species. The critical level of light necessary for shoot production was that a shoot should receive ~10% that of an open sky for *Q. gliva* and ~30% for *Q. acuta*, but both the numbers of shoots produced and shoot length at higher light intensities were much greater for *Q. acuta*, i.e. there is a difference in growth requirement.

8.3.2 Morphogenetic plasticity

The phytochrome family of plant photoreceptors (Taiz et al. 2022, section 16.1) is sensitive to the change in R:FR ratio and can affect transcription regulation. Although R:FR ratio and phytochromes are important, differences in flux density also affect plasticity. Such differences can be detected by two classes of UV-A/blue-light receptors: cryptochromes and phototropins (Vandenbussche et al. 2005; de Wit et al. 2016). Much research into the effects of changing R:FR has been with 'sun' plants, annuals that typically live in environments with full sunshine, and their typical response to a decrease in R:FR is an increase in height growth—frequently interpreted as a character that reduces the effects of competition. Other responses have been found such as elongation of petioles, reduced area of leaves, and decreased chlorophyll (Franklin 2008). These types of plasticity may affect yield. In high-density stands of sunflower (*Helianthus annuus*)—10 and 14 plants m^{-2} as opposed to the more usual 5 of commercial plantings—neighbouring plants inclined to alternative sides of planting rows (Pereira et al. 2017). Yield was reduced when this process was mechanically constrained.

In *Arabidopsis thaliana*, a rosette plant in which the stem internodes between leaves do not expand, an increase in leaf uprightness and greater petiole elongation has been found as a response (Bongers et al. 2018). Stoll and Bergius (2005) used distance statistics to compare the development of two genotypes of *A. thaliana* at different initial spacing patterns: one a wild type (*WT*) and one a transgenic phytochrome A overexpressor (*trans*) in which the normal response to low R:FR was counteracted. Although there was substantial mortality over five weeks in both genotypes it was greater for *trans* (56% mortality) than *WT* (37%). Both random initial sowing patterns and clumped patterns were used. Final patterns of *trans* were regular if the initial pattern was random. Where clumps had been used these were still apparent in *WT* plantings but had disappeared almost completely in *trans*. Interestingly, *WT* populations showed regular patterns of the plant apices at very local distances before the onset of density-dependent mortality even though sowing had been random. Stoll and Bergius attribute this to bending of the hypercotyl on germination due to an early effect of the phytochrome response.

Three processes may modulate the effect of R:FR on increasing height increment of shaded plants. First, a positive response in height increment of, say, partly shaded plants may result in them reaching a height at which the R:FR is greater and so the stimulus, if resulting from detection at the apex, is likely to be reduced. Second, the indication that closer spacing increases the overall height of a stand, seen in the comparison between 2- and 4-cm *T. patula*, suggests that dominant plants as well as subdominants may respond to changes in R:FR that can be sensed in horizontally reflected light (Ballaré et al. 1987). In this sense all plants may respond to shade but none 'avoids' its effects. Third, the stimulus effect of a decrease in R:FR can require a specific level of overall light level being received and may not occur in deep shade. Casal et al. (1986) show this effect on tillering of grass species using supplementary red light at the base of canopies grown at different plant spacing. They suggest 'tillering might be reduced by a phytochrome response triggered by low R/FR before a serious scarcity in energy availability is produced by mutual shading'.

8.3.3 Simultaneous occurrence of multiple types of plasticity

Within tree canopies morphological changes in foliage, shoot, and branch structure occur in gradients from top to the canopy base, but these changes differ according to the particular characteristics of the species. *Thuja plicata* (Cupressaceae) can live for 800–2,000 years (Minore 1990) and is generally considered as shade tolerant (see Edelstein and Ford 2003 for references) and can regenerate under a mature canopy and grow to a mature tree. Edelin (1977) defines *T. plicata* as having Attim's architectural model.

Thuja plicata has a bifurcating branch system with multiple sequential orders of branching (Fig. 8.5a). The adaxial surfaces of small scale-like leaves are adpressed to the shoot (Fig. 8.5b) and the disposition of foliage is determined by the branching process. Edlin (1977) considered that lower (more proximal) order axes, I and II, are axes of exploration of the environment while the higher order axes, III through V, form a frond-like organization. Order IV axes are alternate along order III but order V support foliage within the frond structure with only more distal axes being alternate (Fig. 8.5c).

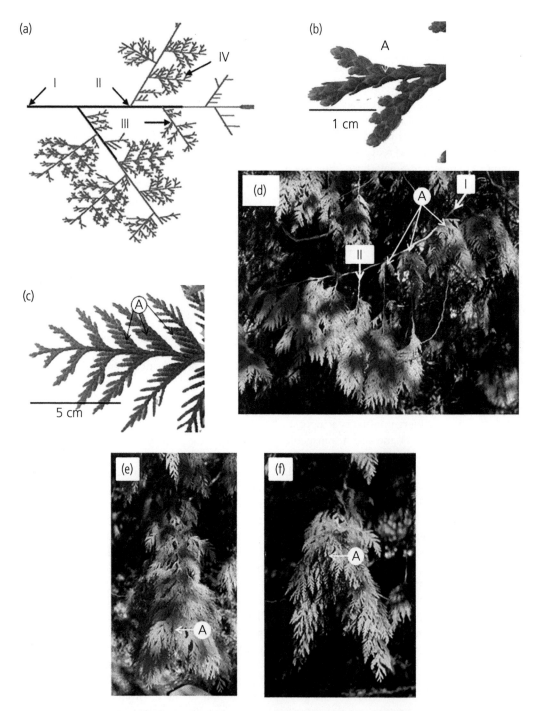

Fig. 8.5 Branching structures of *Thuja plicata*. (a) Diagrammatic representation of the sequence of branching orders. Arrows point to the base of a representative of an order. Not to scale. (b) Apex of a frond, order III, showing scale-like leaves (A) and alternate lateral branches. (c) Distal section of a frond: main axis order III with alternate laterals of order IV. Note that the proximal order V axes (label A) are not alternate but together fill the space between successive order IV axes. (d) A branch, axis I, with the nearside in well-lit conditions but the far-side towards the tree crown interior in shade. Fronds develop from near-side axes II but on the far-side are no longer present. Epicormic branches occur at the base of order II branches that have lost proximal fronds. (e) A branch axis, order I, growing radially out from the crown from above showing alternate frond bearing branches of order II and its apex labelled A. (f) The same branch as in (e) but looking from the apex towards the trunk, illustrating fronds arrayed on both side of the axis. Part (a) reproduced from Edelstein and Ford (2003) with permission from Oxford University Press

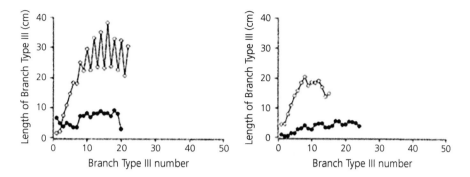

Fig. 8.6 Variation in the structure of fronds in *Thuja plicata*. Two comparisons of branches, all from the same tree, from upper crown (filled circles) and lower crown (open circles) of the lengths of successive order III axes along their order II parents. Left graph for Type II branches at 8th position from the tip of the parent branch Type I; right graph for Type II branches at the 4th position. Upper canopy lengths are shorter, always <10 cm; lower crown order III axes show growth, i.e. increase in length up to nine or ten positions from the apex. The lower canopy branch in the left figure shows cycling of attained length with shorter axes on the more shaded side of the branch. Reproduced from Edelstein and Ford (2003) with permission from Oxford University Press.

The resulting structure of the branch is related to the light conditions in which it grows. On a branch with one side well illuminated but the other facing the interior of the crown and so being more shaded, fronds on the illuminated side develop but those towards the interior may not (Fig. 8.5d), which can be considered the result of resource acquisition plasticity. For a branch developing towards a more even distribution of light on either side, fronds are more equal (Fig. 8.5e, f).

In old-growth *T. plicata* trees ~47 m tall both the distances of fronds along their parent branches (order III along order II) and size of fronds themselves were greater on branches from the lower (at least 20 m from the tree top) part of the canopy compared to those from the top (within 2–3 m) (Edelstein and Ford 2003). Axes within fronds were longer in the lower canopy (Fig. 8.6) and branches were longer. On some lower canopy branches there is cycling of lengths of type II axes along axes II with axes on the more illuminated side being longer, and there was also cycling of branch angles with the side of the branch, with greater lengths having greater declination from the vertical. There was a larger foliage area per unit foliage weight (specific foliage area, SFA) for all branch orders in the lower canopy.

An important feature of some shaded branches, particularly those that have their apex in the crown interior, is that the angle of the order I branch changes from being declined from the horizontal to being vertical, likely the result of morphogenetic

plasticity. This results in the branch being curved (Fig. 8.7a, b). Order II branches being produced at the vertically oriented apex have a spiral arrangement and tend to grow more rapidly than the order I apex. This forms what Edelstein and Ford term 'mini-trees' (their Fig. 10) and produce what Hallé et al. term a reiteration of the tree form. Van Pelt (2001) suggests that old-growth *T. plicata* are prone to loss of the main axis through drought and so reiteration axes succeed it (Fig. 8.7c). In open-grown trees reiterations can occur from close to the ground and can result in a bush-like plant (Edelin 1977). Plasticity is an integral component of the growth of *T. plicata*: it is a dynamic response to growth in shaded conditions within the forest canopy and is manifest in a number of characteristics.

8.4 Changes in plasticity as a canopy grows

In response to changes in planting density the cotton plant (*Gossypium hirsutum*) shows both resource acquisition and morphogenetic plasticity. The plant is shrub-like, growing to ~2 m (Fig. 8.8). In the United States it is typically cultivated at 3.3 plants m^{-2} but this can range between 1.5 and 8.7 plants m^{-2} (Adams et al. 2019). Plants are typically grown in rows ~1 m apart, with variation in density achieved by varying spacing along the rows. Planting density can affect weed suppression, yield quality, timing of yield, and harvesting procedures.

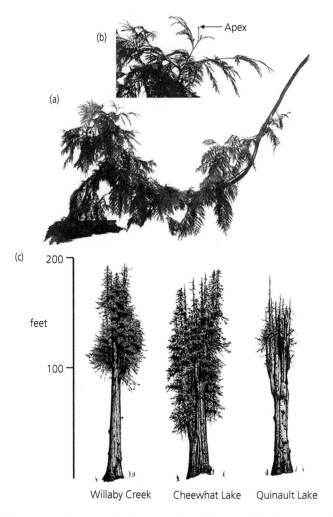

Fig. 8.7 (a) A branch from the lower crown of *Thuja plicata* joined at the trunk on the right and showing a downward and then upward curvature. (b) Enlargement of distal section of the same branch as (a), showing the vertical apex and laterals that extend more rapidly than the apex. (c) Three examples of old-growth *T. plicata* trees mapped and drawn, each showing multiple reiterations: Willaby Creek (Olympic Peninsula) estimated at ~600 years; Cheewhat Lake, estimated as the largest tree in Canada; Quinault Lake tree with a hollow base and small amount of foliage. Part (c) reproduced from Van Pelt (2001) with permission from University of Washington Press.

Architecture of cotton follows Petit's model (Hallé et al. 1978). The central stem is a monopodium and forms through continuous orthotropic growth with leaves spirally arranged and potentially with additional monopodial branches arising from low positions on the main trunk (Fig. 8.8a, b). Each leaf on the monopodium subtends a lateral branch that is sympodial in growth form, with each of its modules having a small proximal scale-like leaf, a foliage leaf separated from it by a long internode and a single terminal flower (Fig. 8.8c, d). The foliage leaf is inserted just below the flower and subtends the next module of the sympodium and leaves are generally horizontal. The resulting bush-like plant is typically ~2 m tall with multiple internal flowers/fruits ranged along its lateral branches.

Leaf and internode plasticity in cotton shows two phases (Dauzat et al. 2008): first, increasing organ size, in both leaf area and internode length, followed

Fig. 8.8 *Gossypium hirsutum* (cotton). (a, b) Whole young plant; (c, d) lateral branch comprised of successive modules. Intervals on measuring stick are 10 cm. Parts (a) and (c) reproduced from Hallé et al. (1978) with permission from Springer.

by decrease in size of both. Dauzat et al. measured seasonal development of cotton plant structure for plants at 1, 2, and 4 plants m^{-2} and calculated leaf irradiance and R:FR at different heights in the canopy which grew ~1.5 m over 120 days following emergence. Plants at wider spacing had more monopodial branches at the early growth stage and greater rates of node appearance on both main stem and sympodial upper branches. Both features can be considered as resource acquisition plasticity.

The transition for internode length on the main stem occurred around leaf node rank 11 at 4 plants m^{-2} but around leaf node 14 at 1 and 2 plants m^{-2}. Node length was greater at 4 than 1 or 2 plants m^{-2} in the first phase but less in the second phase as internode length declined (Fig. 8.9) and during this phase the area of individual leaves produced was less at 4 plants m^{-2} spacing.

The contrasts between plants grown at different planting densities in size of leaf produced and internode length on the first module of sympodial branches was even more marked than differences along the monopodium (Fig. 8.9). The first internode on sympodia originating from nodes 6–11 of

the main stem was twice as long at planting density of 4 compared to 1 plants m^{-2}, but for those from internodes 15 onwards internode length was less at 4 plants m^{-2}.

Dauzat et al. suggest that during the first phase of growth the greater internode lengths on plants at closer spacing were a response to lower R:FR but that in the second phase potential photomorphogenetic responses of internodes were counteracted by plant carbon limitations.

Dauzat et al. note that when internodes are initiated they are at the top of the canopy and the R:FR does not vary greatly between plants growing at different densities. However, the duration of internode elongation is some three weeks and over this period R:FR decreases as growth occurs above the internode and more so for the 4 plants m^{-2} than other planting densities. Sympodia start their development when the main stem has developed three or four modules above their branching node, which may explain why the planting density effect is more pronounced on the growth of sympodial than monopodial stem internodes. Dauzat et al. note that photomorphogenetic responses such as

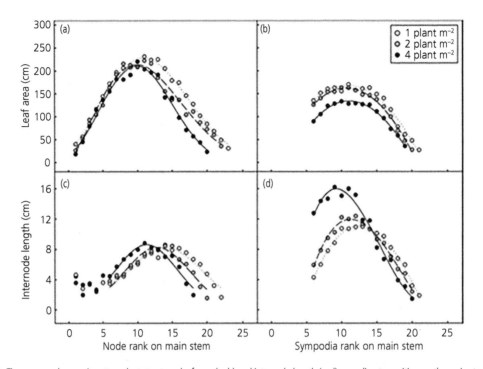

Fig. 8.9 Time-course changes in cotton plant structure: leaf area (a, b) and internode length (c, d) according to position on the main stem (a and c, respectively) and changes in leaf area and internode length on the first module of sympodial branches according to their rank on the main stem (b and d, respectively). Plant densities are indicated on the graphs. Reproduced from Dauzat et al. (2008) with permission from Oxford University Press.

increased internode length 'can only be expressed within the limits allowed by assimilate availability'. Plants may be short of assimilate at later stages. They also found a very marked effect on branch orientation which was isotropic at a plant spacing of 1×1 m but oriented towards the space between rows at greater planting densities.

8.5 Plasticity, canopy light interception, and productivity

Smaller more erect foliage is related to greater yield of more recent maize hybrids (Chapter 2), and Duvick et al. (2004) suggest that more upright leaves improves light utilization at the higher planting densities that are used with later hybrids. Two maize hybrids are described in Chapter 2, one from the 1960s (Hy1960) and one from the 1990s (Hy1990) that has more upright (Fig. 2.7) and smaller leaves. Both hybrids show plasticity in foliage angle with

leaves being more upright at closer initial spacing but this plasticity is greater for Hy1990.

In a spacing trial using two planting densities, 64,000 (LD) and 95,095 (HD) plants ha^{-1}, grain yield per hectare of Hy1990 was greater than Hy1960 at both LD and HD. L was greater at HD for both hybrids (Fig. 8.10) but with the less-productive hybrid, Hy1960, having significantly greater foliage amount than Hy1990 at HD: at day 195, L = 7.40 > 6.41, p <0.05, while at day 211 at LD, L = 5.54 > 4.34, p <0.05. There were no significant differences in plant height between hybrids or planting densities. At day 195 there was significant difference in mean plant total weight between planting densities with HD 73% of LD, but no significant differences between hybrids at each planting density. However, there were significant differences in mean leaf area per plant both between planting densities, with HD < LD, and between hybrids at both densities, with Hy1990 < Hy1960. At HD mean leaf area per plant of Hy1990 was 88% that of Hy1960; at LD it was 90%.

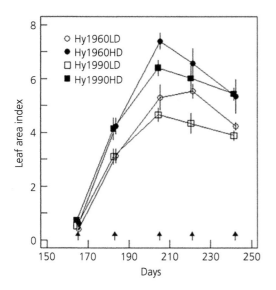

Fig. 8.10 Leaf area index for two maize hybrids grown at 64,000 (LD) and 95,095 (HD) plants ha^{-1} through the growing season. Sample days are indicated by arrows. At both planting densities L for Hy1960 (circles) was greater than that of Hy1990 (squares) during the period of grain filling from day 195.

The ear grew entirely following anthesis, ~ day 195, and there were no significant differences in mean stem weights between days 195 and 232 for any planting density cross hybrid combination, suggesting that differences in ear weight are due to differences in canopy productivity over the 195- to 232-day period, rather than hybrids having differences in 'allocation' between stem and ear (Cliquet et al. 1990).

The decrease of mean plant ear dry weight at HD was greater for Hy1960 (42% less) than for Hy1990 (27% less). This difference is such that for Hy1990 ear weight per hectare was greater at HD whereas for Hy1960 it was greater at LD. Interestingly there was no significant difference between hybrids in mean plant stem weight at either density, although when the hybrids are taken together, mean plant stem dry weight at HD was significantly less than LD. The 1960s hybrid produces more foliage but produces less grain yield and has a greater reduction in grain yield at the higher planting density.

Transmission of light through each canopy was measured at each of the sample times shown in Fig. 8.10 in conditions of direct sunshine and within

two hours of local noon and then calculated as proportional transmission, i.e. below canopy/above canopy. As might be expected, LD treatments that had less L had significantly greater (p <0.05) transmission than HD treatments for both hybrids, e.g. at day 174 (Hy1990, 0.08797 > 0.02447; Hy1960, 0.03567 > 0.02062) and within LD Hy1990 > HY1960, i.e. the canopy of Hy1960 intercepted more radiation. The proportion of light transmitted decreased through to the fourth sample, day 212, for all treatments, with values for LD being significantly greater than those of HD. For the second through fourth samples Hy1990 had a consistently greater transmission than Hy1960. So, while Hy1990 had *greater grain yield it absorbed less light* than Hy1960.

A crucial difference between hybrids was in plasticity in response to changes in planting density: foliage uprightness increased more in response to increased planting density for Hy1990 than Hy1960. The effect of this greater plasticity is seen in the frequency distributions of leaf area at different declinations from the vertical for leaf above the ear at LD and HD (Fig. 8.11). The angles at the mid-point of the cumulative frequency distributions of area by declination changed from LD to HD from 32 to 21° for Hy1990 but only 47 to 44° for HY1960, i.e. Hy1990 shows greater plasticity towards more upright foliage.

All frequency distributions of declination angle are right skewed and the differences between them are in the range of the right tail, i.e. how far it extends towards the horizontal (90° declination) and downward drooping (>90°), and the relative amounts of area in those classes. The maximum amount of leaf area in a 10° declination class is at a declination ≤40°, with the greatest difference between hybrids for leaf ≥13 (Fig. 8.11). For leaf ≥13 Hy1990 not only has more upright foliage, as expected, but also the decrease in declination induced by closer spacing is greater than for Hy1960.

The plasticity in leaf declination for Hy1990 that results in smaller declination angles produces differences in patterns of leaf illumination. A canopy can be considered more efficient as the amount of its foliage receiving light between the light compensation point (LCP) and the light saturation point (LSP) is increased. The shape of the photosynthesis curve

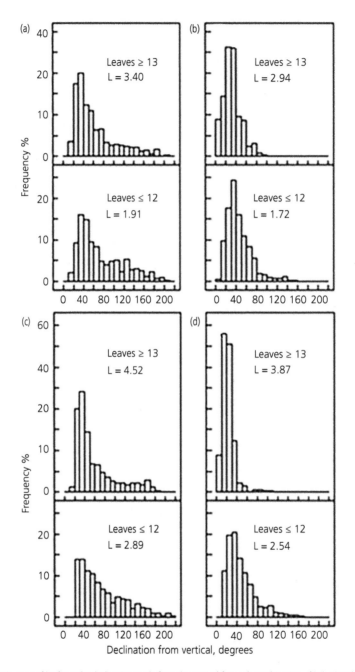

Fig. 8.11 Frequency distributions of leaf area by declination angle from the vertical for each combination of hybrid and planting density: (a) Hy1990 HD; (b) Hy1990 LD; (c) Hy1960 HD; and (d) Hy1960 LD. Histograms were calculated by pooling values for 5-cm leaf length sections for three intensively sampled plants within each combination of hybrid and density treatment at the day 195 sample. At the planting densities used in this work a single ear is produced from the axil of leaf 12; on most plants the maximum individual leaf area was at position 13 and the canopy is considered in two parts: that for leaf 13 and above ($\geq$13) and that for leaf 12 and below ($\leq$12).

in maize varies with leaf position (e.g. Moreno-Sotomayor et al. 2002), age (e.g. Thiagarajah et al. 1981; Usuda 1984; Dwyer and Stewart 1986; Stirling et al. 1994), and leaf nitrogen content and genotype (e.g. Settimi and Maranville 1998). Grain filling occurs when the canopy is generally stable, during which time the topmost and youngest leaves age to some 40 days and the larger and older leaves to more than that. Moreno-Sotomayor et al. (2002; Fig. 2) found an LCP of ~100 and LSP of ~1,100 μmol s^{-1} m^{-2} typical for an aged leaf 13 of hybrid Hy1990 planted at 64,000 plants ha^{-1}. For a value of above canopy illumination of 1,800 μmol s^{-1} m^{-2}, typically found during measurements over two hours on either side of solar noon, PAR, between 100 and 1,100 μmol s^{-1} m^{-2} are equivalent to 5–60% of above canopy. Even should there be no clear saturation in the photosynthesis curve, as shown by Moreno-Sotomayor et al. (2002; Fig. 3) for leaf 17 compared to leaf 13 in

late-July (Nebraska), there is a curvilinear decreasing photosynthetic rate in relation to PPFD so that equalizing light interception values away from values both below the compensation point and high values would increase overall efficiency in canopy photosynthesis.

For all combinations of hybrid and planting density upper canopy leaves intercept light at a higher percentage of the above canopy value than lower canopy leaves and with an extended range of percent illumination relative to above canopy values (Fig. 8.12). However, for leaf ≤12 only Hy1990 LD shows substantial illumination at greater than 5% of the above canopy value and interestingly both Hy1960 LD and Hy1990 HD have substantial areas at <5% and <10% but Hy1990 has a greater total percentage of area receiving illumination between 10 and 50% of above canopy values.

Leaf ≥13 of Hy1990 HD had substantially greater L in the declination interval 10–30° than Hy1960 but

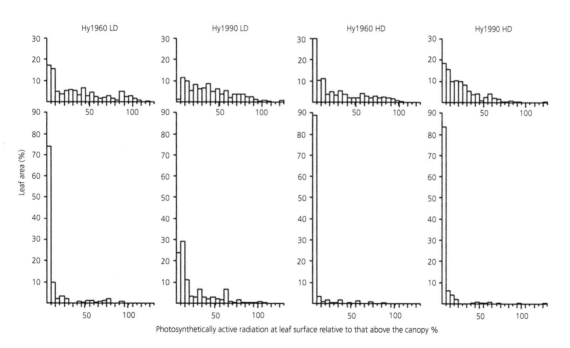

Fig. 8.12 Instantaneous measurements of illumination relative to that received on a horizontal plane above the canopy, measured within two hours of solar noon, of 5-cm-long leaf sections at the day 195 sample, grouped in 5% increments of the proportion of above canopy, horizontal-position value for each hybrid and planting-density combination. Both adaxial and abaxial leaf surfaces were measured and the greater value between them used, but the abaxial surface rarely received more than 5% of above canopy values and rarely (<1%) received greater photon flux density than the adaxial surface. Upper graphs for leaves ≥13, lower graphs for leaves ≤12. Leaf sections angled so that they intercept light normal to the sun's rays have values greater than 100% of above canopy light as measured on a horizontal surface.

less in the declination interval 30–80° (Fig. 8.11). For leaf ≤12 the values for HD are greater for all declination angles (Fig. 8.11), reflecting the substantial difference in total L between HD and LD. The greater penetration of light to leaf ≤12 of LD compared to HD results in a greater leaf area intercepting at all values above 5%. Most leaf ≤12 of HD is in substantial shade, reflecting the large L of leaf ≥13. The greater production of HD relative to LD is likely a result of greater photosynthate production in the upper canopy.

For Hy1990 LD 76.5% of leaf ≥13 received light between the assumed LCP and LSP compared to 60.2% for Hy1960 LD. For leaf ≤12 at LD the difference is greater, with 70.5% Hy1990 receiving light between LCP and LSP compared to 21% for Hy1960. At LD the more upright leaved hybrid Hy1990 had markedly greater light penetration to leaf ≤12.

At the higher planting density the pattern of light distribution changed for both hybrids but in different ways. For Hy1960 HD 52.4% of leaf ≥13 received light between LCP and LSP, a small decrease from LD although a greater decrease than for Hy1990. For Hy1990 the overall distribution is different between LD and HD: at LD there is only 1% of leaf area below LCP but 22.5% above LSP; for HD there is 18% below LCP and 7.1% above LSP. Hy1990 achieves a greater area of foliage between LCP and LSP at HD compared to LD through a marked plasticity of increase in foliage uprightness in leaf ≥13.

More upright foliage can increase canopy efficiency in conversion of light to yield and this effect does not depend upon there being L ≥5. The greater canopy efficiency of the more upright-leafed hybrid outweighed the effect of greater light absorption by the more horizontal leafed hybrid with a greater estimated foliage area between compensation point and saturation point but this was achieved in two ways. At the lower planting density of 64,000 plant ha^{-1} it was sufficient to produce increased illumination to lower canopy leaves; at the higher planting density of 95,095, where L was greater, leaves became more upright in the upper canopy and this increased foliage area between compensation point and light saturation point. Sarlikioti et al. (2011) stress the importance of calculating effects of variation in leaf angle on light interception and photosynthesis in canopies of tomato (*Solanum lycopersdicum*) and Niinemets (2010) reviews the effects of variation in foliage angle along with other foliage properties in relation to canopy photosynthesis and variation in shade tolerance.

8.6 Conclusion

A strength of the Hallé et al. theory is that it identifies architectures common to species with different geneologies—but information is required that enables understanding of how species that may have substantial differences nevertheless function with the same architectural model. Hallé et al. repeat that architectural models are first identified through study of isolated plants and Barthélémy and Caraglio (2007) suggest the same approach. Their viewpoint is that models are expressed clearly in such plants. The idea of there being a default standard architecture for a species expressed when a plant is grown in isolation discounts two essential aspects of plant growth:

- through plasticity growth can result in different structures and growth amounts, depending upon conditions; and
- that there is variation between individuals that may enable growth in different conditions.

Although Hallé et al. make a distinction between endogenous effects, i.e. the basic architectural model, and exogenous effects, environmentally induced plasticity, they stress the importance of reiteration in a number of models, which in a number of examples is the result of plastic development.

Plasticity is not found only in plants that have been over-topped; it determines how dominant plants develop. Few plants habitually grow isolated from neighbours, although bristlecone pine may be one, in common with other plants of desert or semi-desert habitats. Even the foliage of *Arabidopsis thaliana*, a short-lived rosette species, shows plasticity in its foliage. An understanding of how architecture influences growth dynamics requires integration with foliage structure and distribution.

The productivity of maize hybrids illustrates effects of plasticity and variation. Group selection for yield per area of ground rather than yield per plant resulted in smaller, more upright leaves and consequently an improved light distribution across foliage for plant photosynthesis and accentuation of these effects through plasticity under close spacing.

The variation that can be seen within species, or in the case of maize between hybrids, combined with the fact that plasticity may occur for all plants in a stand suggests that *developmental tolerance* is an essential property for plants that habitually grow in closed canopies. This tolerance to variation in environmental conditions is achieved through genetic variation, so that some proportion of individuals may grow and reproduce under different conditions, and plasticity in a range of characters that influences how architectural models function as conditions vary during growth.

8.7 References

Adams, C., Thapa, S. and Kimura, E. 2019. Determination of a plant population density threshold for optimizing cotton lint yield: a synthesis. *Field Crops Research*, 230, 11–16.

Ballaré, C. L., Sanchez, R. A., Scopel, A. L., Casal, J. J. and Ghersa, C. M. 1987. Early detection of neighbour plants by phytochrome perception of spectral changes in reflected sunlight. *Plant, Cell and Environment*, 10, 551–557.

Ballaré, C. L., Scopel, A., Jordan, E. and Vierstra, R. D. 1994. Signalling among neighboring plants and the development of size inequalities in plant populations. *Proceedings of the National Academy of Sciences of the United States of America*, 91, 10094–10098.

Barthélémy, D. and Caraglio, Y. 2007. Plant architecture: a dynamic, multilevel and comprehensive approach to plant form, structure and ontogeny. *Annals of Botany*, 99, 375–407.

Bednarz, C. W., Nichols, R. L. and Brown, S. M. 2006. Plant density modifications of cotton within-boll yield components. *Crop Science*, 46, 2076–2080.

Black, J. N. 1964. An analysis of the potential production of swards of subterranean clover (*Trifolium subterraneum* L.) at Adelaide, South Australia. *Journal of Applied Ecology*, 1, 3–18.

Bongers, F. J., Pierik, R., Anten, N. P. R. and Evers, J. B. 2018. Subtle variation in shade avoidance responses may have profound consequences for plant competitiveness. *Annals of Botany*, 121, 863–873.

Brisson, J. 2001. Neighborhood competition and crown asymmetry in *Acer saccharum*. *Canadian Journal of Forest Research*, 31, 2151–2159.

Casal, J. J., Sanchez, R. A. and Deregibus, V. A. 1986. The effect of plant density on tillering: the involvement of R/FR ratio and the proportion of radiation intercepted per plant. *Environmental and Experimental Botany*, 26, 365–371.

Cliquet, J.-B., Deléens, E. and Mariotti, A. 1990. C and N mobilization from stalk and leaves during kernel filling by ^{13}C and ^{15}N tracing in *Zea mays* L. *Plant Physiology*, 94, 1547–1553.

Dauzat, J., Clouvel, P., Luquet, D. and Martin, P. 2008. Using virtual plants to analyse the light-foraging efficiency of a low-density cotton crop. *Annals of Botany*, 101, 1153–1166.

De Wit, M., Keuskamp, D. H., Bongers, F. J., et al. 2016. Integration of phytochrome and cryptochrome signals determines plant growth during competition for light. *Current Biology*, 26, 3320–3326.

Duvick, D. N., Smith, J. S. C. and Cooper, M. 2004. Long-term selection in a commercial hybrid maize breeding program. *Plant Breeding Reviews*, 24, 109–151.

Dwyer, L. M. and Stewart, D. W. 1986. Effect of leaf age and position on net photosynthetic rates in maize (*Zea mays* L.). *Agricultural and Forest Meteorology*, 37, 29–46.

Edelin, C. 1977. Images de l'architecture des Conifères. *Thèse. Doctoral. 3^0 Cycle, Université. Montpellier II*.

Edelstein, Z. R. and Ford, E. D. 2003. Branch and foliage morphological plasticity in old-growth *Thuja plicata*. *Tree Physiology*, 23, 649–662.

Ford, E. D. 1975. Competition and stand structure in some even-aged plant monocultures. *Journal of Ecology*, 63, 311–333.

Ford, E. D. 2014. The dynamic relationship between plant architecture and competition. *Frontiers in Plant Science*, 5, Article 275.

Franklin, K. A. 2008. Shade avoidance. *New Phytologist*, 179, 930–944.

Hallé, F., Oldeman, R. A. A. and Tomlinson, P. B. 1978. *Tropical Trees and Forests: An Architectural Analysis*. New York, Springer.

Kenkel, N. C. 1988. Pattern of self-thinning in Jack pine: testing the random mortality hypothesis. *Ecology*, 69, 1017–1024.

Koike, F. 1989. Foliage-crown development and interaction in *Quercus gilva* and *Q. acuta*. *Journal of Ecology*, 77, 92–111.

Minore, D. 1990. *Thuja plicata* Donn ex D. Don Western Redcedar. In: R. M. Burns, R. M. and B. H. Honkla, B. H. (eds) *Silvics of North America*. Volume I Conifers. Washington, DC, Forest Service Department of Agriculture. http://www.na.fs.fed.us/spfo/pubs/silvics_manual/volume_1/thuja/plicata.htm.

Moreno-Sotomayor, A., Weiss, A., Paparozzi, E. T. and Arkebauer, T. J. 2002. Stability of leaf anatomy and light response curves of field grown maize as a function of age and nitrogen status. *Journal of Plant Physiology*, 159, 819–826.

Morgan, D. C. and Smith, H. 1978. The function of phytochrome in the natural environment. VII. The relationship between phytochrome photo-equilibrium and development in light grown *Chenopodium album* L. *Planta*, 142, 187–193.

Nagashima, H. 1999. The process of height-rank determination among individuals and neighbourhood effects in *Chenopodium album* L. stands. *Annals of Botany*, 83, 501–507.

Nagashima, H. and Terashima, I. 1995. Relationships between height, diameter and weight distributions of *Chenopodium album* plants in stands: effects of dimension and allometry. *Annals of Botany*, 75, 181–188.

Nagashima, H., Terashima, I. and Katoh, S. 1995. Effects of plant density on frequency distributions of plant height in *Chenopodium album* stands: analysis based on continuous monitoring of height-growth of individual plants. *Annals of Botany*, 75, 173–189.

Niinemets, U. 2010. A review of light interception in plant stands from leaf to canopy in different plant functional types and in species with varying shade tolerance. *Ecological Research*, 25, 693–714.

Pereira, M. L., Sadras, V. O., Batista, W., Casal, J. J. and Hall, A. J. 2017. Light-mediated self-organization of sunflower stands increases oil yield in the field. *Proceedings of the National Academy of Sciences of the United States of America*, 114, 7975–7980.

Sarlikioti, V., De Visser, P. H. B., Buck-Sorlin, G. H. and Marcelis, L. F. M. 2011. How plant architecture affects light absorption and photosynthesis in tomato: towards an ideotype for plant architecture using a functional–structural plant model. *Annals of Botany*, 108, 1065–1073.

Settimi, J. R. and Maranville, J. W. 1998. Carbon dioxide assimilation efficiency of maize leaves under nitrogen stress at different stages of plant development. *Communications in Soil Science and Plant Analysis*, 29, 777–792.

Sinclair, T. R. and Muchow, R. C. 1999. Radiation use efficiency. *Advances in Agronomy*, 65, 215–265.

Smith, H. and Whitelam, G. C. 1997. The shade avoidance syndrome: multiple responses mediated by multiple phytochromes. *Plant, Cell and Environment*, 20, 840–844.

Stirling, C. M., Aguilera, C., Baker, N. R. and Long, S. P. 1994. Changes in the photosynthetic light response curve during leaf development of field-grown maize with implications for modeling canopy photosynthesis. *Photosynthesis Research*, 42, 217–225.

Stoll, P. and Bergius, E. 2005. Pattern and process: competition causes regular spacing of individuals within plant populations. *Journal of Ecology*, 93, 395–403.

Taiz, L., Zeiger, E., Møller, I. M. and Murphy, A. 2022. *Plant Physiology and Development*, Seventh Edition. Oxford, Oxford University Press.

Thiagarajah, M. R., Hunt, L. A. and Mahon, J. D. 1981. Effects of position and age on leaf photosynthesis in corn (*Zea mays*). *Canadian Journal of Botany*, 59, 28–33.

Turley, M. C. and Ford, E. D. 2011. Detecting bimodality in plant size distributions and its significance for stand development and competition. *Oecologia*, 167, 991–1003.

Umeki, K. 1995. Modeling the relationship between the asymmetry in crown display and local environment. *Ecological Modelling*, 82, 11–20.

Umeki, K. and Seino, T. 2003. Growth of first-order branches in *Betula platyphylla* saplings as related to the age, position, size angle and light availability of branches. *Canadian Journal of Forest Research*, 33, 1276–1286.

Uria-Diez, J. and Pommerening, A. 2017. Crown plasticity in Scots pine (*Pinus sylvestris* L.) as a strategy of adaptation to competition and environmental factors. *Ecological Modelling*, 356, 117–126.

Usuda, H. 1984. Variations in the photosynthesis rate and activity of photosynthetic enzymes in maize leaf tissues of different ages. *Plant and Cell Physiology*, 25, 1297–1301.

Van Pelt, R. 2001. *Forest Giants of the Pacific Coast*, Seattle, University of WashingtonPress.

Vandenbussche, F., Pierik, R., Millenaar, F. F., Voesenek, L. A. C. J. and Van Der Straeten, D. 2005. Reaching out of the shade. *Current Opinion in Plant Biology*, 8, 462–468.

Van Pelt, R. 2001. *Forest Giants of the Pacific Coast*. Seattle, University of Washington Press.

Weller, D. E. 1987. A reevaluation of the −3/2 power rule of plant self-thinning. *Ecological Monographs*, 57, 23–43.

PART IV

Growth response to environmental change

Introduction to Part IV

The environment varies continuously—with regular patterns in some variables, such as day length, and at least partial irregularity for others, such as those associated with weather systems—and variations in what we might consider as environmental variables are always correlated. Environmental conditions can affect different component processes of plant growth, but change in conditions is inevitable and growth processes both are adapted to regular variation and may acclimate to less regular changes that have important effects on growth.

Seasonal variation in weather is ubiquitous—even in the tropics—and in many regions variation in the weather follows oscillations, which though not as regular as those of the day–night cycle, and with different characteristic amplitudes of variation, induce responses in plant processes. Plant control systems are integrated with this variation, which in some cases initiates a physiological or developmental process and sometimes controls rate.

Chapter 9 provides examples of the effects of these different patterns of variation in a plant's external environment and discusses the physiological acclimation to them that produces particular phases of growth and the domains of an environmental condition when its impact on growth may be strong. There can be changes in the environment that a plant can acclimate to and maintain growth and conditions, but some conditions may be experienced when such acclimation may be unable to compensate for changes. An example of shoot growth measured hourly over a period of 21 days illustrates different types of control systems with different characteristics in both in the delay between a change in the environment and response of the plant as well as in the amplitude of responses. Some of this variation is due to interactions between control systems such as those influencing developmental and autotrophic processes. An example is given of control systems that may have limits to their effectiveness.

As plants grow their internal environment changes. An important example occurs in the process of plant ageing, discussed in Chapter 10 using examples of semelparous plants that have terminal sexual reproduction and iteroparous plants that have continuous sexual reproduction as they grow and frequently grow to a large size. In both types of plant changes in the internal environment control the success of reproduction and longevity of the plant. Understanding these changes and their effects requires integrating knowledge of morphological, and sometimes anatomical, development, with that of physiology—physiology alone does not provide an explanation for the changes that take place in growth as plants age. In both types of plant the ageing processes vary between species and the characteristic environments of the plants.

CHAPTER 9

Growth in a fluctuating environment

9.1 Introduction

Plant growth in the variable environment is the result of interaction between two systems: the physiological and morphological system of the plant and the physical–chemical system of the environment. These have different dynamic properties and examples are given of time scales, amplitudes of variation, and interactions. Plants are adapted both to utilize and at least partly tolerate cycles of variation in the physical–chemical system, but their responses to change are not constant—the same environmental change can produce different responses at different times—but controlled by the genetics of the plant.

The natural environment is not constant. Conditions affecting the component processes of plant growth rate fluctuate at different time scales: diurnal, in response to weather systems lasting a few days, and seasonal. These fluctuations have components of both regularity and irregularity—particularly where weather patterns are concerned. And there are interactions between variables, for example an increase in incoming radiation during a day is likely to increase air temperature.

Shoot extension of a tree, *Picea sitchensis*, provides an example of growth responses to changes in the environment. Analyses of shoot growth have used an *environment* → *growth* framework and attempted to define extension as a response to environmental factors using various types of correlation analyses between weather variables and increment. Not surprisingly, studies using this approach have resulted in different conclusions about relationships between environment and growth. This chapter shows it is more appropriate to analyse the control processes involved and to define the domains of influence of environmental factors.

9.2 Empirical studies of shoot extension rate

Field measurements of shoot extension for a range of species have shown that environmental variables may interact in affecting extension rates. For *Picea abies* in New Hampshire, USA, and Ontario, Canada, Worrall (1973) found a positive effect of increased mean daily temperature (degrees Centigrade) on daily shoot growth (y, centimetres) but that this effect was greater on moist days. He summarized this relationship in a regression equation, using mean daily relative humidity (RH) and humidity deficit (HD) as the difference in RH between 06:00 and 12:00 h for the day minus the same value for the previous day: $y = 2.86 + 0.017T + 0.0082RH + 0.0096HD$. Worrall suggests that apparently conflicting published results over whether growth was greater at night or during the day may be resolved by considering effects of humidity variables and different species responses to them.

Working with daily measurements of extension of the leading shoot of Sitka spruce, y_t, Ford et al. (1987a) used time series analysis to investigate the relationship with environmental variables. They found lagged relationships with both mean daily temperature, T_t, and daily total solar radiation, R_t, summarized in the form:

$$y_t = 0.133T_{t-1} - 0.042T_{t-2} + 0.0107R_{t-2} + 0.0150R_{t-3}.$$

This suggests daily increment is positively related to the previous day temperature, t_{-1}, but negatively related to the day before—reflecting the cyclical pattern of temperature variation that occurred during

The Dynamics of Plant Growth. E. David Ford, Oxford University Press. © E. David Ford (2023). DOI: 10.1093/oso/9780192867179.003.0009

the measurement period; extension is also positively related to the total radiation of the two and three days previously, as indicated by the last two terms on the right-hand side of the equation. However, solving this equation first required that a seasonal trend of gradual increase then decrease (Ford et al. 1987b) be removed so that these environmental effects are to be considered within a seasonal framework. Consequently this equation only applied to the first part of annual growth when temperature variation was substantial and had a distinct pattern.

Measurements made over the course of a day show that shoots may both extend and contract at distinct times in the diurnal cycle. Kanninen (1985) found an influence of temperature on shoot extension of *Pinus sylvestris* in southern Finland but with a lag of two hours in its effect and suggested that temperature operates through a cycle of shoot contraction due to water loss and extension due to rehydration.

Bermann and DeJong (1997a, 1997b) note that a diverse range of plants including grasses and conifers as well as a number of *Prunus* species, which they measured, show maximum growth rates between 16:00 and 19:00 h during periods of rapid water potential recovery. Shoot extension rate of *Prunus persica* in the field was lowest early in the morning and increased throughout the day with a two- to three-fold increase in late afternoon sustained for two to four hours, and after this surge extension declined markedly and remained low during the night. Bermann and DeJong (1997b) determined temperature response of extension at high water potential in growth room studies but found field measurements deviated from this temperature response proportional to the decrease in shoot water potential. The field relationship between shoot extension and shoot water potential may not be simple: shoots of *P. persica* in an experimental water stress treatment decreased in length in the morning, as expected, yet resumed growth in the afternoon even when shoot water potential was lower than it had been in the morning when shoot extension decreased—water-stressed plants can adjust cell wall properties to regulate growth when turgor changes, an indication of an active control process.

Attempts to provide explanations of variation in shoot extension in terms of variation in physical variables—to seek mechanistic explanations—have not provided definitive answers but have given rise to questions such as: How is it that plants appear to respond differently to environmental variations at different times? How can shoot extension be restricted with a decrease in shoot water potential but restart at a lower shoot water potential than appeared to have initiated the restriction? Answers to such questions require analysis of the dynamic properties of plants and their coupling with the environment.

9.3 Overview of dynamic processes influencing shoot extension

Analysis of dynamics requires recognizing that multiple component systems each having particular characteristics and interactions with other components can contribute to shoot extension. Three selected components are represented in Fig. 9.1:

1. Developmental physiology. This determines when growth can occur, and in some cases may set limits to growth rate. In this chapter the focus is on seasonal and diurnal cycles.
2. Physical environment. Diurnal and seasonal variations in environmental factors influence photosynthesis, evapotranspiration, and growth.
3. Plant water relations. Inflow of water into new cells is an active process that drives their expansion and so the growth of new tissue. Water loss causes reduction in tissue water potential that can affect cell water uptake and consequently both construction and expansion of existing cells and new cell growth. Water is lost from plant tissues through transpiration and replaced by root uptake, so the time course of plant water potential can be affected by characteristics of foliage, wood, and roots. Continued evapotranspiration from the stand over successive days can decrease soil water potential if there is no replenishment by precipitation, which in turn may affect the characteristics of diurnal variation in tissue water potential.

There are two requirements when considering these components and their interactions:

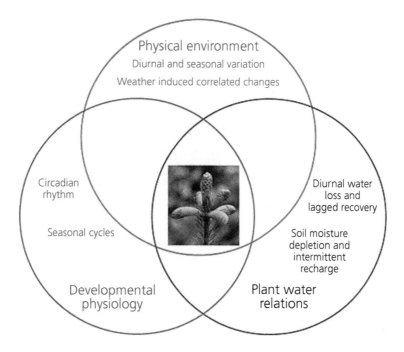

Fig. 9.1 Shoot extension is affected by three constituent systems. Each has dynamic components; two examples that can influence the course of extension over time are given for each system. Central image of *Picea sitchensis* shoots shortly after bud break with bud scales still over some of the developing shoots. Dreamstime.com images.

1. Define domains of influence. A domain defines the conditions under which particular influences may operate. This can be under regulatory control by the plant. For example, developmental processes restrict shoot growth to a particular period which defines the domain of influence of environmental conditions.

2. Define effects of change over time. An environmental change might stimulate or depress growth but such responses are unlikely to be instantaneous, so their full effects require definition over a time period that may include considering a compensatory, or a recovery, response. The example is given of water loss through transpiration slowing shoot extension, or even causing shrinkage, with time taken for recovery.

A technique in analysis of dynamic systems is to make parallel measurements of changes of both inputs to the system, in this case weather variables and measures of soil water, over sustained periods of time. An example of this approach is given for the shoot extension of trees in a young plantation of Sitka spruce (*Picea sitchensis*) in south-west Scotland.

Auxanometer studies measuring shoot extension (e.g. Idle 1956) have been made by attaching a sensing device to the growing shoot but the attachment may have to be reset at intervals. In order not to interfere with the processes of extension and contraction, and to extend uninterrupted measurements over a number of days, an instrument was constructed to scan the position of the growing apex without touching it (Milne et al. 1977) (Fig. 9.2a). This used short pulses of infrared radiation at 900 ± 50 nm, which is outside the action spectra of plants and measured extension of the leading shoot of a tree, which continues the main axis of the tree vertically (Fig. 9.2b). The instrument was fixed to the previous year's hardened woody growth and could be used continuously over an extended period.

Hourly values of shoot extension (Fig. 9.3) shown for 10 days illustrate development of a diurnal

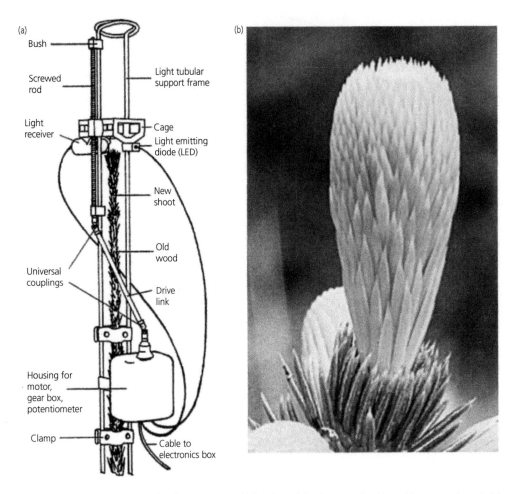

Fig. 9.2 (a) Shoot extension sensor mounted on the previous year thickened wood showing sensor head in position to scan the end of the growing tip of the new extension; at each scan time the motor drives the head upwards until the beam is not blocked and then down until intercepted and the distance is measured with a potentiometer. (b) Young growing shoot of *P. sitchensis* showing the flat top formed by needle points. Part (a) reproduced from Milne et al. (1977) with permission from Wiley. Part (b) photo credit: geograp.org.uk.

cycle during a period of transition from overcast or partially overcast days (24–26 June) to generally clear conditions (up to 3 July). There is gradual development of reduced extension during the middle of the day (25–29 June), with small contractions on some days. As clear conditions continue there are substantial mid-day contractions (30 June–2 July), which were also found for days following 3 July (not shown). Three other shoots showed the same pattern of hourly extension.

9.4 Seasonal patterns of shoot extension and developmental physiology

Daily shoot extension totals were calculated for the 24-h period from immediately pre-dawn so that any contraction in a day, the influence of which is discussed later, was counted along with subsequent night-time recovery. Although there are differences in total growth between shoots from different trees (Fig. 9.4b), the general features of

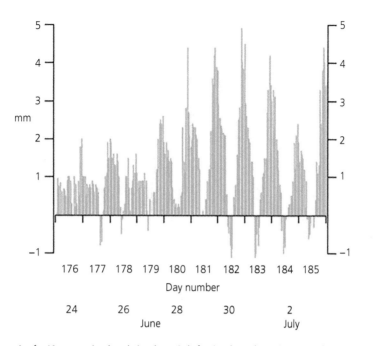

Fig. 9.3 Hourly shoot extension for 10 consecutive days during the period of active shoot elongation as weather transitioned from overcast (24–26 June) to clear conditions.

day-to-day variation are similar, for example for each shoot there is a general increase in daily extension up to 29 June followed by a decrease. Some local maxima, notably 21 June, and minima, notably 2 July, are also seen in all three shoots. The trees were of different sizes and made different total amounts of extension over the 21 day period: shoot 1, 748 mm cm (red in Fig. 9.4b); shoot 2, 552 mm (green); shoot 3, 490 mm (blue). Each of these trees had its leading shoot and top whorl of branches above the interlocking branches of the canopy.

The general increase of the extension made in 24-h periods up to 29 June (Fig. 9.4) followed by a reduction is a 'seasonal' effect (Ford et al. 1987a) which is an integral component of the annual growth cycle in trees involving release from dormancy in late winter, bud burst in spring, active growth cessation in late summer, followed by winter dormancy (Fig. 9.5) (Singh et al. 2017). While the general increase in extension occurs during a period of increase in mean daily air temperature and daily total radiation (19–29 June) (Fig. 9.4a) the decrease following 29 June takes place while temperature does not decrease and radiation decreases

but then returns to close to its previous maximum value.

The circadian clock (Chapter 5) controls the annual cycle of development and seasonality of growth. It is entrained to ambient day length with a phytochrome-mediated process being involved. Singh et al. (2017) summarize a transcriptional network for photoperiodic control of shoot growth and its transition to bud development in poplar. Expression of the *CONSTANS* (*CO*) gene is controlled by the circadian clock to peak at the end of the daily light period. CO targets expression of the *FLOWER-ING LOCUS* (*FT2*) gene and a cascade to cell cycle genes (Fig. 9.6), but the CO protein is unstable in the dark so that *FT* expression is weak under short days. Short days induce the removal of growth-promoting signals rather than generating a growth inhibitor. *FD-LIKE1* (*FDL1*) and *ABSCISIC ACID INSENSITIVE 3* (*ABI3*) are involved in regulation of bud maturation and cold acclimation.

Cooke et al. (2012) review coordination by the circadian clock of plant hormone metabolism and sensitivity of plants to hormones and note that a reduced level of gibberellic acid (GA) is required

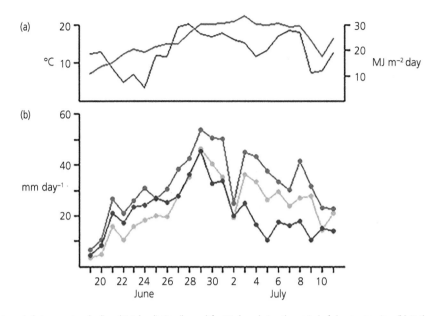

Fig. 9.4 (a) Mean daily temperature (red) and total radiation (brown) for 21 days during the period of shoot extension. (b) Daily extension of the leading shoots of three *P. sitchensis* trees over the same period following bud break: shoot 1 red; shoot 2 green; shoot 3 blue. For both (a) and (b) values are from pre-dawn (05:00 h) of the indicated day to 04:00 h of the next. Hourly values over 10 days for shoot 2 are shown in Fig. 9.3.

for growth cessation and that short days downregulate GA20 oxidase, the enzyme for the rate-limiting step in GA biosynthesis, and suggest that reduction of active GA rather than sensitivity towards GA is important in growth cessation. Abscisic acid (ABA) affects the circadian clock and the clock controls plant sensitivity changes through the day. Cooke et al. suggest that the precise role of ABA in bud development, dormancy, and maturation is not clear.

There are differences between species in this control system. In *Picea abies* expression of an *FT*-like gene, *FT4*, is induced by onset of short days and its expression results in induction of growth cessation (Gyllenstand et al. 2007). Closely related *FT* genes are known to have opposite effects (e.g. Lee et al. 2013).

An important component of growth control involves movement of molecules into the bud and within the shoot apex. Cell-to-cell signalling is a dominant feature of growth and maintenance of activity of the shoot apical meristem (SAM) (Chapter 5). van der Schoot and Rinne (2011) highlight the importance of plasmodesmata in this movement, suggesting (Rinne and van der Schoot

1998) that cell zonation in the SAM is achieved through establishment of functioning connections between cells of the same group while connections between distinct zones are blocked. Blocking is through callose deposition at the plasmodesmatal opening.

Pore exclusion size of symplastic connection within the SAM is reduced shortly after short day (SD) treatment and van der Schoot and Rinne (2011) propose that this decreases transport capacity. Cell walls are also modified. Symplastic isolation and cell wall modification correspond with dormancy, but during release of chilling-induced dormancy symplastic connections are restored by callose degradation. Cooke et al. (2012) suggest: 'it is possible that plasmodesmatal connectivity is required to enable growth *per se* rather than mediate dormancy transitions'. GA amount, controlled by the circadian clock, is implicated in this process.

Day length, and so control via the circadian clock, is implicated in both bud break and shoot growth cessation, with the method of control linked to growth-promoting signals. Experimental day length manipulation has led to the concept of critical day length (CDL) for growth cessation, which in *P.*

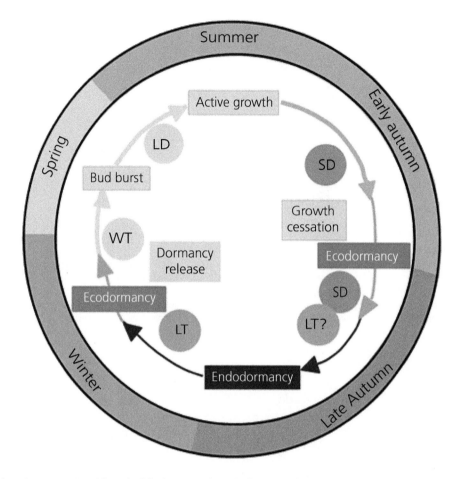

Fig. 9.5 Schematic representation of the cycle of development and growth of tree shoots based principally on research into poplar and *Picea*. LT low temperature; WT warm temperature; LD long days; SD short days. SD during late summer/early autumn induces the cessation of growth and bud development. The first stage of growth cessation is called ecodormancy, during which cessation can be reversed by exposing plants to growth-promotive conditions. Prolonged exposure to SDs in late autumn induces the establishment of dormancy in buds. The plant is in an endodormant state when it became insensitive to growth-promotive signals. Low non-lethal temperatures produce release from endodormancy, following which growth arrest is maintained by external signals (LT) and buds are described as being in another ecodormant state. Increasing temperatures, a few degrees higher than the endodormancy-releasing temperature, promote spring bud burst. In some species such as birch and apple, LT induces both dormancy establishment and release and promotes bud burst. Reproduced from Singh et al. (2017) with permission from Wiley.

abies may be affected by temperature (e.g. Kvaalen and Johnsen 2008).

Maximum shoot elongation (Fig. 9.4b) occurs on 29 June (day-of-year 181, eight days after the summer solstice). For another year Ford et al. (1987b) report a maximum for leading shoots on day 180 and shoots lower in the canopy having earlier times of maximum growth, varying from day 174 for leading shoots of the topmost whorl to day

170 for shoots at whorl 5 below the leading shoot (Fig. 9.7). Generally, shoots lower in the canopy started and ceased growth earlier than those close to the top. The leading shoot and shoots of topmost whorl branches of some trees had a distinct secondary maximum of growth (Fig. 9.7). Ford et al. (1987a) note: 'The differences which accounted for the largest part of variation in total shoot length were the duration and rate of growth in the later

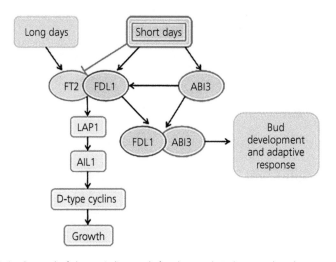

Fig. 9.6 Underlying transcriptional network of photoperiodic control of poplar growth. A change to short days represses *FT2* expression and results in downregulation of *LIKE-AP1 (LAP1)* and the *LAP1* target *AINTEGUMENTA-LIKE1 (AIL1)*. Downregulation of *AIL1* leads to suppression of cell-cycle-related genes and growth cessation. Expression of *FD-LIKE1 (FDL1)* and *ABA INSENSITIVE 3 (ABI3)* is induced by short days, and they form a transcriptional complex involved in regulation of bud maturation and cold acclimation-related actions. Reproduced from Singh et al. (2017) "Photoperiod- and temperature-mediated control of phenology in trees—a molecular perspective". *New Phytologist* 213(2):511–524, with permission from Wiley.

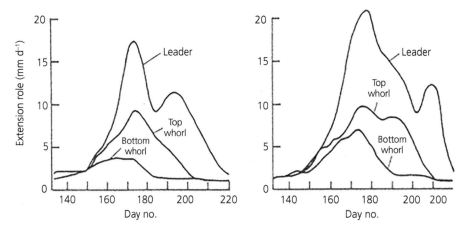

Fig. 9.7 Twenty-one-point moving averages of daily shoot increment on two trees of *P. sitchensis* for the leading shoot, i.e. continuation of the trunk, and the topmost whorl and the 5th whorl from the top labelled as the bottom whorl. Reproduced from Ford et al. (1987a) with permission from Oxford University Press.

part of the season, i.e. after the time of maximum trend in increment.'

Studies of circadian clock-based photoperiodic control of shoot growth in trees have concentrated on the events of bud burst and bud formation and winter hardening. However, Cooke et al. (2012) note 'that the state of dormancy exhibited by the SAM during the activity–dormancy cycle is quantitative rather than absolute'. If this applies through the growth period then it may contribute to an explanation of the increase then decrease in shoot extension between bud break and bud formation. The changes that may occur in symplastic transport capacity could also contribute to a trend in growth rate.

Whilst the details of seasonal growth control have not been established, the existence of distinct quantitative seasonality is clear and the well-established role of the circadian clock in control of bud formation and in some species of bud break and suggests a role in growth seasonality. The circadian clock coupled with phytochrome and possibly other diurnally fluctuating components provides active control by the biological system.

Interestingly, the period of daily maximum shoot growth is after the longest day, suggesting a lagged effect in control. For uppermost shoots there is a distinct secondary maximum in growth, suggesting conditions that may counter ecodormancy.

9.5 Patterns of environmental variables

Annual and diurnal cycles dominate patterns of variation in physical variables. These are primarily driven by solar radiation but modified by energy exchange processes that produce lagged effects and modify amplitudes of responding conditions. While radiation is received by plants in a diurnal cyclical pattern, most frequently there are differences in quantities from day to day (Fig. 9.8). The amount of visible light received and its effects on photosynthesis are frequently studied, but visible light is just one component of total radiation. Radiation is a determinant of conditions such as temperature and evapotranspiration that in turn affect growth.

Temperature and the difference between daily maxima and minima depend upon the particular weather system—in the example (Fig. 9.8) there is transition from a period of low pressure with overcast skies to a high pressure system with clearer skies. The temperature range is greater on sunny days, although minimum temperature did increase and was greater than on cloudy days, with the exception of early morning of 27 June.

Environmental conditions are partially correlated in time. For example, total radiation reached distinct diurnal maxima for days 27 June through 3 July (Fig. 9.8) and maximum air temperature consistently followed maximum total radiation by between two and three hours; air temperature reached a greater maximum on sunny days. This is likely due to the processes whereby radiation causes a temperature increase of the forest biomass and the heat involved is subsequently transferred to the air through re-radiation, advective and convective cooling,

and evapotranspiration (Monteith and Unsworth 2007).

9.5.1 Transpiration

When viewed as controlled by the physical properties of a system, such as heat capacity and heat conductivity and physical aspects of cooling, the energy exchange process exerts passive control. However, stomatal (g_s) and mesophyll (g_m) conductance are both controlled by botanical processes, which implies active control. Both may be expected to vary over time and so, possibly, affect plant water potential and thus growth. However, plant water potential is affected not only by water loss, through transpiration, but also by both water uptake and storage in the plant (Taiz et al. 2022, Chapter 6).

Evaporation occurs directly from moist surfaces and through transpiration from within plants. Both processes obviously have a cooling effect determined by the latent heat of evaporation, but transpiration rate is affected by the greater pathway of resistance from within foliage, i.e. the combined effect of mesophyll resistance and stomatal resistance (Evans and Loreto 2000; Flexas et al. 2008). Heating and cooling of air involves transfers of energy, and in the case of evapotranspiration of mass, and the rate of transfer contributes to the observed delay between maximum radiation and maximum temperature. This delay is less on days with less radiation, 24 June and 9 July. In the diurnal sequences air temperature declines from its maximum to a minimum, which for days from 27 June onwards occurred around dawn the following morning when radiation was received again.

Water lost through transpiration can cause a reduction in plant water potential which in turn can affect growth; transpiration rate depends upon both physical conditions of the atmosphere and properties of the plant and canopy. The Penman–Monteith equation (Monteith 1965) provides a definition of this system and is used extensively to estimate transpiration, particularly where there is a need to estimate water loss from a crop to calculate irrigation requirements (e.g. Allen et al. 1998, but see also Allen et al. 2020). It incorporates two terms: g_c, canopy conductance, the rate of water vapour movement out of the plant which is affected by

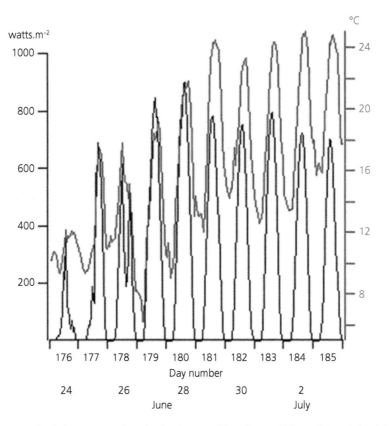

Fig. 9.8 Total radiation received and air temperature above the closed canopy of the ~15-year-old forest of *Picea sitchensis* for a period of 10 days during transition from cloudy (days 175 and 176) to generally clear conditions (days 180–184)—the same days as shown in Fig. 9.3 for hourly shoot extension.

mesophyll and stomatal resistance; and g_a, the boundary layer conductance, the rate at which water vapour moves from the canopy space to the atmosphere above it. The Penman–Monteith equation may be written thus:

$$E_t = \frac{sA + c_p\rho Dg_a}{\lambda\{s + \gamma(1 + g_a/g_c)\}}$$

where E is mass water evapotranspiration rate (g s^{-1} m^{-2}), D is saturation vapour pressure deficit of the air, A is available energy, and s is the slope of the relation between saturation vapour pressure and temperature. λ, c_p, ρ, and γ are, respectively, the latent heat of vaporization of water, the specific heat of air at constant pressure, the density of air, and the psychrometric constant, and these parameters vary with air temperature. The numerator term defines the energy available for transpiration

and the dependence of its use on air conditions, notably saturation vapour pressure deficit, which varies with temperature. On days of greater radiation the maximum calculated transpiration can be delayed some two hours after maximum radiation, which reflects its dependence upon conditions of the atmosphere. The denominator defines application of the latent heat of vaporization used to calculate mass of water from available energy and operates in relation to the psychrometric constant which relates the partial pressure of water in the air to air temperature.

A difficulty in using the Penman–Monteith equation lies in estimating a value for canopy conductance. Milne et al. (1985) consider the canopy of the Sitka spruce forest, where the measurements of shoot extension were made, as composed of a number of categories of the foliage canopy within which stomatal conductance could be expected to be similar. They estimated a value using

$$g_c = \sum_{i=1}^{N} (g_{si}L_i)$$

where g_{si} is stomatal conductance of shoots in the ith of N categories and L is the foliage area index in the ith category. Stomatal conductance was measured hourly for 10 shoots (Leverenz et al. 1982) using a 'null balance' continuous flow type diffusion porometers (Beardsell et al. 1972). From the investigations of Milne et al. (1985) and Milne (1989) empirical summary relationships were established between g_c and total radiation and g_a and wind speed. Variability in g_{si} and considerable variance in measured L_i for this forest lead to uncertainty in calculation of E_t using the Penman–Monteith equation, so it is important to consider the effectiveness of this equation. Transpiration was also estimated for this forest on a separate occasion using an energy balance equation applied to successive 20-minute periods.

Estimates of transpiration calculated by the two methods were significantly correlated and the two methods show the same general diurnal pattern. However, they did not have a 1:1 relationship, indicating some bias in one or other of the estimates. Milne et al. concluded that an important contributing difficulty was obtaining satisfactory values of g_{si} and L_i for use in calculating canopy conductance in the Penman–Monteith equation. Use of the Penman–Monteith equation for calculating estimates of transpiration should be taken as an indication for comparison between days rather than a definitive estimate of transpiration amount.

For this *P. sitchensis* canopy, under conditions when there was no surface moisture on the foliage, transpiration was an important contributor to energy transfer, counting for between 40 and 60%.

9.6 Plant water relations and their effects on extension and growth

Shoot water potential depends upon the difference between water lost from the plant through transpiration and gained through movement from stem, trunk, and roots. Transpiration in the Sitka spruce

forest could remain high well into the afternoon due to the lag in air temperature relative to the radiation maximum.

Water potential of young foliated shoots was measured (Milne et al. 1983) using a Scholander-type pressure bomb (Scholander et al. 1965). Days of parallel data for estimated transpiration and measured shoot water potential, for the same year as the shoot measurement sequence shown in Fig. 9.4 but after shoot extension had ceased, are shown in Fig. 9.9 (for an extended series see Milne et al. 1983). Water is first lost from foliage and as this loss occurs on clear days water potential of the foliage declines rapidly as transpiration increases. Then towards night, as transpiration declines to close to zero, shoot water potential increases. This increase depends upon water movement along the gradient of potentials from soil to atmosphere, referred to as the soil–plant–atmosphere continuum, and its rate depends upon the resistance to water flow in plant tissues and size of the difference in water potentials and, as recovery proceeds, then the difference declines. Shoot water potential tends to reach maximum values late at night, before dawn, with values measured in the Sitka spruce trees at that time generally between −0.2 and −0.5 MPa. There is a lag between the decline in transpiration and increase in shoot water potential, but by ~05:00 h shoot water potential is at ~93–96% of its pre-dawn maximum of ~0.32 MPa (Fig. 9.9).

9.7 Diurnal patterns of shoot extension

9.7.1 Cell growth

Plant cell growth takes place while the cytoplasm is enclosed by a flexible primary cell wall that enables extension; subsequent to extension a thicker secondary cell wall is formed. Growth of the primary cell wall requires carbon compounds and it is driven by active uptake of water that generates a turgor pressure. This wall contains 15–40% cellulose, 20–30% xyloglucans, and 35–50% pectin, with small amounts of structural protein and there is considerable heterogeneity in chemistry and structure within and between species (Burton et al. 2010). The primary cell wall has viscoelastic properties: it can be load bearing in the solid phase and

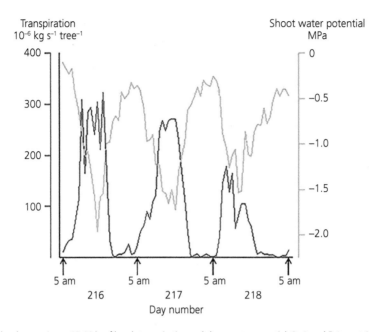

Fig. 9.9 Three diurnal cycles, starting at 05:00 h, of hourly transpiration and shoot water potential: 3, 4, and 5 August 1976. Hourly transpiration from a Sitka spruce forest, estimated by solving the heat balance equation, and shoot water potential measured with a Scholander-type pressure bomb. Reproduced from Milne et al. (1983) "Time lags in the water relations of Sitka spruce". *Forest Ecology and Management* 5(1):1–25, with permission from Wiley.

extensible in the liquid phase. Peaucelle et al. (2012) suggest that two integrated load bearing components are involved in growth and the solid↔liquid transition of the primary cell wall: (i) hemicelluloses (e.g. xyloglucans; Cosgrove and Jarvis 2012). Acid can be released from the cytoplasm into the wall that will stimulate growth and (ii) pectate molecules are bound together with calcium ions and these bonds can be distorted by the wall tension produced by turgor pressure. New pectate deposits can bind at these sites and so material is deposited both on and within the primary cell wall (Boyer 2009).

When cells expand their increase in volume is mostly due to water uptake (Taiz et al. 2022, section 5.6, and reviews by Geitmann and Ortega 2009 and Kutschera and Niklas 2013). For this to happen there must be a water potential difference between the inside (lower potential) and outside (higher potential) of the cell. The growing cell must be maintained at a lower water potential than that in the adjacent xylem and ground tissue. Two processes may contribute to this:

1. Osmotic adjustment: maintenance of a pH difference across the tonoplast results in solutes, cations, and sugars being pumped from the cytosol through the tonoplast into the cell vacuole. Water moves through the plasma membrane, then the tonoplast, into the vacuole through aquaporins which are highly regulated small membrane proteins (Chaumont and Tyerman 2014).

2. The volume adjustment of the growing cell itself. The primary cell wall of the growing cell stretches, a process known as *stress relaxation* (McQueen-Mason et al. 1995). Wall growth occurs, which increases cell volume, lowers turgor pressure and water potential, and can contribute to maintaining the water potential difference between the inside and outside of the cell and the capacity to continue growth (Westgate and Boyer 1984; Boyer 2009).

Measurements of water potential, such as those shown in Fig. 9.9, are made on whole shoots. The extending cells in the meristematic tissue are a small component of such shoots and it is their water potential that must be maintained at a lower value than the body of the whole shoot itself, so that through osmotic adjustment and stress relaxation water moves into the growing cells.

Following the expansion phase of cell growth the secondary cell wall is deposited comprising mainly cellulose, which increases the rigidity of the wall, and then, in woody tissue, cell death including autolysis of cell contents.

9.7.2 Contraction and re-expansion of shoot tissue

The contraction seen in the hourly sequence of measurements in days with a clear sky can be attributed to a rapid decline in shoot water potential that may reduce cell wall growth in meristematic and developing cells and cause elastic shrinkage in cells with recently formed primary cell walls, i.e. where cell wall thickening is not complete. Proseus et al. (1999) show that both these processes can occur in maize leaves but their actions are difficult to separate in growing cells. The recovery of extension following this contraction due to a physiologically induced increase in cell turgor through osmotic adjustment would be a slower process than the earlier reduction in water potential of the whole shoot as it would require activation of enzyme systems involved in solute accumulation in cell vacuoles and possible synthesis of aquaporins and their incorporation into plasma membrane and tonoplast (Chaumont and Tyerman 2014).

The difference between rates of contraction and re-expansion and growth contributes to a diurnal cyclical relationship between estimated hourly transpiration and shoot extension (Fig. 9.10a). For both of the two days shown in Fig. 9.10 values immediately pre-dawn, i.e. before estimated transpiration had started, were close to an extension of 2 mm h^{-1}. As daylight hours progress and transpiration increases, hourly shoot extension declines, reaching ~0 for 29 June but ~−0.5 mm for three successive hours for 3 July, i.e. continuous shrinkage. Estimated transpiration from the middle of the day through to early evening is greater for 3 July than 29 June and hourly shoot extension is less.

Cyclical progression of the relationship between hourly values of shoot extension and estimated transpiration is an example of *hysteresis* (lagging behind) in which the state of the system depends upon its history—the same value of hourly transpiration has different values for shoot extension depending upon the water status of the plant for the particular day and its effect on the cycle of water loss and recharge. Two processes may contribute to this lag: (i) water is lost from foliage and the growing shoot (by transpiration) more rapidly than it is taken up (by movement along the gradient of potentials from soil to foliage and active uptake into cells); and (ii) the rate of cell extension depends upon cell turgor pressure and may increase gradually as water is regained by cells. The period of rapid increase in shoot extension started at 14:00 h for 29 June but 19:00 h for 3 July (Fig. 9.10a).

Following the last hour with positive estimated transpiration in each cycle (Fig. 9.10b) there is a gradual decline in extension. Generally, over this period, i.e. the five or six hours before dawn, shoot water potential is at a stable diurnal maximum, tending to show only a slight increase, if any change (Fig. 9.9), and temperature changes little—most cooling occurs in late afternoon and early evening when hourly extension is at its highest rates. Possible explanations for this decline in hourly shoot extension during the end of the dark period may be reduction in available photosynthate or circadian clock control.

9.7.3 Phases in the diurnal course of shoot extension

Shoot extension rate over the measured sequence of days (Fig. 9.4) is influenced by the progress of seasonal effects and plant water relations which determine daily contraction, recovery, and growth. The magnitude of the effects of plant water relations directly on extension, and possible influence of additional factors, can be understood through comparison of diurnal progress of shoot extension and environmental variables.

A number of phases can be recognized in the pattern from 05:00 h on 29 June (day 181) to 04:00 h on 30 June (day 182). From 05:00 to 08:00 h (phase α in Fig. 9.11) hourly increment declined from 2.1 to

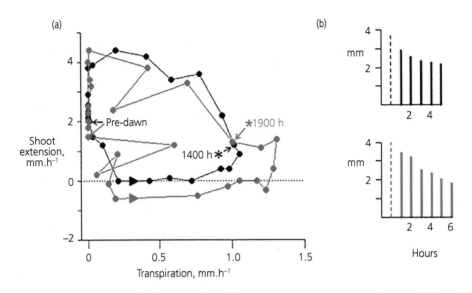

Fig. 9.10 (a) Relationship between estimated hourly transpiration and shoot extension from pre-dawn to pre-dawn the following day: 29 June (day 181), black; 3 July (day 185), red. Pre-dawn values are marked with arrows of respective day colours; times of the start of greatest shoot increment are marked by asterisk and time in hours. (b) Shoot extension for the hours following the last hour with positive transpiration, which is shown as a dashed line.

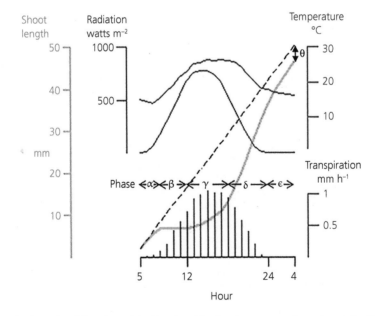

Fig. 9.11 The 24-h course of shoot length from dawn of day 181 when daily extension rate was at its maximum value. The maximum values for air temperature and estimated transpiration lag that of radiation. Five phases of shoot extension rate are identified, α through ε; see text for a description of their characteristics.

1.2 mm and then was zero for the next four hours (phase β), a period of increase in estimated transpiration. Over the following six hours (phase γ) there was a gradual increase to 1.2 mm h^{-1}, occurring during the period of maximum transpiration, followed by a marked increase in rates from 3.6 up to 4.4 mm during the decline of transpiration to zero (phase δ). Finally (phase ε) extension gradually declined from 2.9 to 2.2 mm (see Fig. 9.10b) with similar values to those of the last four hours of the previous 24 h which ended at 04:00 h.

Despite transpiration of only up to 0.5 mm h^{-1} shoot extension was zero during phase β, but as transpiration increased in phase γ nevertheless shoot extension increased; this is consistent with induction of active processes for cell water uptake that maintain turgor but with this induction process causing a lag. Phase δ is the principal recovery period, as transpiration declines, with rates of extension up to twice those of the late-night growth phase, ε. This may be caused by re-expansion of cells that had contracted earlier. This faster extension rate in phase δ may also be due to prior accumulation of cell wall materials in the cytoplasm during the period when no, or reduced, extension occurred. Proseus and Boyer (2005) found in *Chara coralline* that polysaccharides do not enter the cell wall spontaneously but may remain in a layer adjacent to the wall when cell turgor is low. They propose that increased turgor is necessary for wall precursors to be moved from cytoplasm to cell wall.

The dashed black line in Fig. 9.11 marks the extension that would have been achieved had the initial rate of the 24 h been maintained throughout. The difference between final actual extension for the 24 h and this projected amount is 3.85 mm, θ (Fig. 9.11), and is considered as the restriction of growth over the day due to water potential effects. The late night–early morning decline in extension is clearly not related to water potential restrictions but may be related to a decline in available wall substrates.

9.7.4 Variation in the phases of extension

The course of shoot extension for other days is compared with that of 29 June (Fig. 9.12). For both 24 and 25 June more shoot extension occurred than predicted by extrapolation of the pre-dawn rate.

That this occurred for 24 June, when both total radiation and temperatures were lower than the other days shown, suggests that an effect of seasonality to increase growth and similar increases were found for days before 24 June.

Declines in shoot extension during the day are most marked for days of large amounts of total radiation, high temperatures, and large amounts of estimated transpiration (Fig. 9.12; 1, 2, and 9 July). On those days this check to shoot elongation started early in the day; however, on 25 June, when radiation, temperature, and calculated transpiration only markedly increased at 14:00 h the check to shoot elongation was delayed until this time.

The patterns of 1 and 2 July are similar to that of 29 June, although there is actual shrinkage of shoots rather than just cessation of extension. However, on both of these days extension recommences while transpiration is still taking place. The length of time when the shoot contracted was 10 h for 1 July but 12 h for 2 July, although calculated transpiration values for the whole day were almost identical (11.9 mm for 1 July, 11.6 mm for 2 July). A further difference between these two days is that the rates of the post-contraction period are markedly greater for 1 July relative to 2 July for both the rapid period, 20:00–24:00 h (3.5 relative to 2.0 mm h^{-1}), and the subsequent period of slower extension rate, 24:00–04:00 h (2.7 and 1.2 mm h^{-1}). The deficit in extension of 1 July relative to continuation of the pre-dawn rate is 10.5 mm relative to an actual extension of 35.1 mm, whereas for 2 July the comparable values are 21.5 and 19.35 mm, i.e. the apparent reduction in extension due to contraction and cessation of extension early in the day is greater than the actual daily increment for 2 July. The pattern of an accelerated rate of extension later in the day typically between 20:00 and 24:00 h following a slower rate is similar to that reported by Bermann and DeJong (1997a), although it is more delayed and, unlike their example, extension does continue after this accelerated rate, i.e. from 24:00 to 04:00 h.

The recovery of shoot extension from zero or contraction even as estimated transpiration increased suggests an active process of cell turgor regulation. This is obviously lagged and it seems likely becomes limited during extended days of water loss.

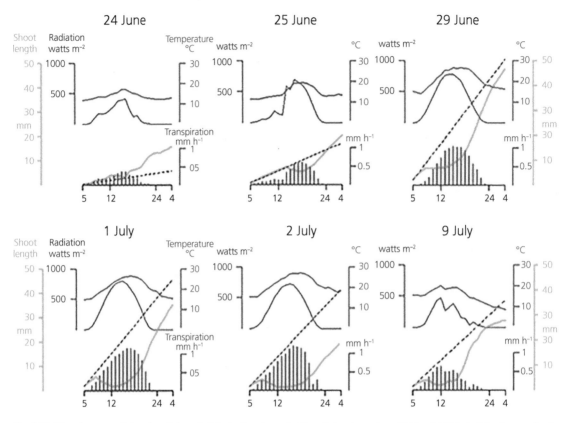

Fig. 9.12 Diurnal patterns of shoot length (green) of the leading shoot of a *Picea sitchensis* tree from 05:00 to 04:00 h the following morning for six selected days (for shoot 2 shown in Fig. 9.4). The black dashed line is a linear projection of shoot length at the extension rate of the first hour. The upper graphs for each day show measured hourly total radiation and mean temperature. The blue bar graph shows hourly values of estimated transpiration.

9.7.5 Uptake of soil moisture

The period 29 June–2 July, during which shoot length contraction increased during the mid-day period, is one of continuing transpiration loss with no overall water gain by the soil. Deans (1979) measured soil moisture, temperature, and fine root amount at this forest site during this year. Measurement of soil moisture was by miniature porous pot tensiometers, with mercury manometers, that could be placed in different horizons of the peaty soil. Fine root growth increased in all horizons during 3–28 May, a period of high rainfall and zero soil moisture tension and increasing temperature, and was generally stable during June. Root amount decreased rapidly, 27 June to 2 July, a period when only 5 mm of rain fell above the canopy. During this period decreasing amounts of fine roots were

significantly correlated with soil moisture tension in each of the top three soil horizons and the decrease occurred first in the surface horizon which had the highest fine root density. Tang and Boyer (2008) report root-mediated water potential limitations to leaf growth in maize.

9.8 Conclusion

Environmental conditions change continuously and although these changes clearly affect growth rate they do not *control* it—the active responses that plants make to these changes determine growth rate. Active controls (Fig. 9.13) produce responses that maintain growth during conditions that may otherwise be considered as restrictive; however, these controls may only function during particular

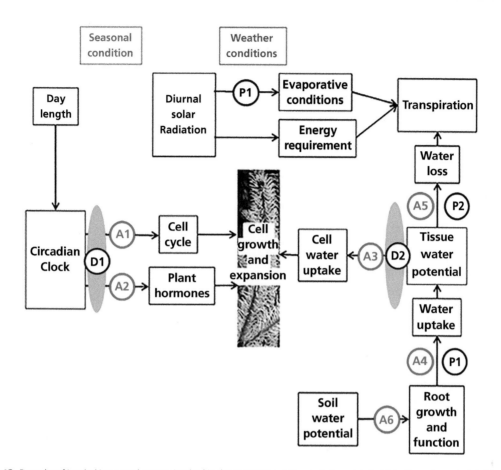

Fig. 9.13 Examples of interlocking control systems involved in shoot extension. Active controls are labelled A (red); passive controls P (blue); domain control D (green). A1 and A2 control synthesis of cell cycle and plant hormone components and are limited to a seasonal domain defined by the day length → circadian clock system. Active uptake of water by expanding cells, A3, operates in the domain of reduced tissue water potential, D2. Water uptake and loss from the plant through transpiration is controlled by both active and passive control processes described in the text.

conditions, i.e. they have domains of operation. Typically the response to changing conditions is not instantaneous and there are conditions when the adjustments have limited effectiveness.

Components of developmental physiology define domains of action for a number of growth processes. There are a number of components of developmental processes with apparently different modes of action but represented by A1 and A2 in Fig. 9.13. Control through day length via the sequence *circadian clock → expression of CONSTANS gene → decay of the gene product in dark* seems to be central, but there are additional components of circadian clock control through regulation of plant

hormone metabolism and plasmodesmata function. Temperature may also affect developmental processes (Chapter 7).

An interesting question is the conditions under which regulatory control through the circadian clock acts as a switch, as in bud formation and winter hardening, or is quantitatively variable, as raised by Cooke et al. (2012), and how such variation is achieved. The concept of ecodormancy, where a dormancy signal can be modified, seems particularly important when considering growth. The occurrence of a secondary maximum of extension for some upper shoots of *Picea sitchensis* trees, which may be related to conditions during ecodormancy, is important since it provides a major contribution to overall shoot length achieved by those shoots. In this case the developmental process is modified.

The cell turgor maintenance process may have a number of components including active control of cell osmotic potential and aquaporin function, A3 in Fig. 9.13. It functions in growth to maintain cell expansion under conditions when water potential of growing tissue declines. Its domain of action, D2, is defined by the tissue water potential that initiates the control actions and the lower tissue water potential at which its control processes are not effective. Of course, tissue water potentials in turn are affected by the balance between transpiration loss and water uptake. There are lags in the cell turgor maintenance process that may be related to activation of osmotic and aquaporin regulation and rates of water movement along water potential gradients within the plant and soil. The components of the cell turgor maintenance process may not respond linearly to changes in water potential and contribute to the hysteresis effect in the transpiration:extension rate relationship (Fig. 9.10). There are conceptual advantages in considering a domain for action of this maintenance process: at water potential greater than the initiating water potential the effects of the turgor control process in terms of lags and differences in rate of action do not apply, as might be considered for the late night periods of shoot extension; at lower water potentials when turgor control does not function growth ceases.

The *environment* → *growth* model for investigation makes the assumption that plants translate differences in environmental factors in a way that can be readily measured as a growth response. It is envirocentric and has been applied with a straightforward view of how plants function. In the example of Sitka spruce shoot extension in a naturally varying environment what we do *not* see is empirical constancy between growth rate and a quantity such as radiation amount or a condition such as temperature.

The empirically based correlation-seeking approach may result in valuable findings where an environmental factor is dominant, for example in cold environments where temperature variation is a dominant factor. However, advances in plant physiology indicate that control of physiological processes, particularly nuclear control, is continuous and results in adjustments that may maintain growth in the face of stress or schedule it to avoid likely unfavourable conditions. This control is additional to that found in acclimation and developmental processes that modify anatomy and morphology, but when taken together knowledge of these processes is sufficient to indicate that when considering plant growth the properties of the plant should be central and considered in relation to how effects of inevitable changes in environment may be controlled.

9.9 References

Allen, R. G., Pereira, L. S., Raes, D. and Smith, M. R. 1998. *Crop Evapotranspiration: Guidelines for Computing Crop Water Requirements*. Rome, United Nations FAO.

Allen, R. G., Dukes, M. D., Snyder, R. L., Kjelgren, R. and Kilic, A. 2020. A review of landscape water requirements using a multicomponent landscape coefficient. *Transactions of the ASABE*, 63, 2039–2058.

Beardsell, M. F., Jarvis, P. G. and Davidson, B. 1972. Null-balance diffusion porometer suitable for use with leaves of many shapes. *Journal of Applied Ecology*, 9, 677–690.

Berman, M. E. and Dejong, T. M. 1997a. Crop load and water stress effects on daily stem growth in peach (*Prunus persica*). *Tree Physiology*, 17, 467–472.

Berman, M. E. and Dejong, T. M. 1997b. Diurnal patterns of stem extension growth in peach *Prunus persica*: temperature and fluctuations in water status determine growth rate. *Physiologia Plantarum*, 100, 361–270.

Boyer, J. S. 2009. Cell wall biosynthesis and the molecular mechanism of plant enlargement. *Functional Plant Biology*, 36, 369–385.

Burton, R. A., Gidley, M. J. and Fincher, G. B. 2010. Heterogeneity in the chemistry, structure and function of plant cell walls. *Nature Chemical Biology*, 6, 724–732.

Chaumont, F. and Tyerman, S. 2014. Aquaporins: highly regulated channels controlling plant water relations. *Plant Physiology*, 164, 1600–1618.

Cooke, J. E. K., Eriksson, M. E. and Junttila, O. 2012. The dynamic nature of bud dormancy in trees: environmental control and molecular mechanisms. *Plant, Cell and Environment*, 35, 1707–1728.

Cosgrove, D. and Jarvis, M. 2012. Comparative structure and biomechanics of plant primary and secondary cell walls. *Frontiers in Plant Science*, 3, Article 204.

Deans, J. D. 1979. Fluctuations of the soil environment and fine root growth in a young Sitka spruce plantation. *Plant and Soil*, 52, 195–208.

Evans, J. R. and Loreto, F. 2000. Acquisition and diffusion of CO_2 in higher plant leaves. in: Leegood, R. C., Sharkey, T. D. and von Caemmerer, S. (eds) *Photosynthesis: Physiology and Metabolism*, pp. 321–351. Dordrecht, Kluwer Academic Publishers.

Flexas, J., Ribas-Carbó, M., Diaz-Espejo, A., Galmés, J. and Medrano, H. 2008. Mesophyll conductance to CO_2: current knowledge and future prospects. *Plant, Cell and Environment*, 31, 602–621.

Ford, E. D., Deans, J. D. and Milne, R. 1987a. Shoot extension in *Picea sitchensis*. 1. Seasonal variation within a forest canopy. *Annals of Botany*, 60, 531–542.

Ford, E. D., Milne, R. and Deans, J. D. 1987b. Shoot extension in *Picea sitchensis*. 2. Analysis of weather influences on daily growth rate. *Annals of Botany*, 60, 543–552.

Geitmann, A. and Ortega, J. K. E. 2009. Mechanics and modeling of plant cell growth. *Trends in Plant Science*, 14, 467–478.

Gyllenstrand, N., Clapham, D., Källman, T. and Lagercrantz, U. 2007. A Norway spruce FLOWERING LOCUS T homolog is implicated in control of growth rhythm in conifers. *Plant Physiology*, 144, 248–257.

Idle, D. B. 1956. Studies in extension growth. 1. A new contact auxanometer. *Journal of Experimental Botany*, 7, 347–361.

Kanninen, M. 1985. Shoot elongation in Scots pine: diurnal variations and response to temperature. *Journal of Experimental Botany*, 36, 1760–1770.

Kutschera, U. and Niklas, K. J. 2013. Cell division and turgor-driven stem elongation in juvenile plants: a synthesis. *Plant Science*, 207, 45–56.

Kvaalen, H. and Johnsen, O. 2008. Timing of bud set in *Picea abies* is regulated by a memory of temperature during zygotic and somatic embryogenesis. *New Phytologist*, 177, 49–59.

Lee, R., Baldwin, S., Kenel, F., McCallum, J. and Macknight, R. 2013. FLOWERING LOCUS T genes control onion bulb formation and flowering. *Nature Communications*, 4, Article 2884.

Leverenz, J., Deans, J. D., Ford, E. D., Jarvis, P. G., Milne, R. and Whitehead, D. 1982. Systematic spatial variation of stomatal conductance in a Sitka spruce plantation. *Journal of Applied Ecology*, 19, 835–851.

McQueen-Mason, S. J. and Cosgrove, D. J. 1995. Expansin mode of action on cell walls—analysis of wall hydrolysis, stress relaxation, and binding. *Plant Physiology*, 107, 87–100.

Milne, R. 1989. Diurnal water storage in the stems of *Picea sitchensis* (Bong.) Carr. *Plant, Cell and Environment*, 12, 63–72.

Milne, R., Smith, S. K. and Ford, E. D. 1977. An automatic system for measuring shoot length in Sitka spruce and other plant species. *Journal of Applied Ecology*, 14, 523–529.

Milne, R., Ford, E. D. and Deans, J. D. 1983. Time lags in the water relations of Sitka spruce. *Forest Ecology and Management*, 5, 1–25.

Milne, R., Deans, J. D., Ford, E. D., Jarvis, P. G., Leverenz, J. W. and Whitehead, D. 1985. A comparison of two methods of estimating transpiration rates from a Sitka spruce plantation. *Boundary-Layer Meteorology*, 32, 155–175.

Milne, R., Smith, S. K. and Ford, E. D. 1977. An automatic system for measuring shoot length in Sitka spruce and other plant species. *Journal of Applied Ecology*, 14, 523–529.

Monteith, J. L. 1965. Evaporation and environment. *Symposia of the Society for Experimental Biology*, 19, 205–234.

Monteith, J. L. and Unsworth, M. H. 2007. *Principles of Environmental Physics*, Third Edition. New York, Academic Press.

Peaucelle, A., Braybrook, S. and Hofte, H. 2012. Cell wall mechanics and growth control in plants: the role of pectins revisited. *Frontiers in Plant Science*, 3, Article 121.

Proseus, T. E. and Boyer, J. S. 2005. Turgor pressure moves polysaccharides into growing cell walls of *Chara corallina*. *Annals of Botany*, 95, 967–979.

Proseus, T. E., Ortega, K. E. and Boyer, J. S. 1999. Separating growth from elastic deformation during cell enlargemant. *Plant Physiology*, 119, 775–784.

Rinne, P. L. H. and van deer Schoot, C. 1998. Symplasmic fields in the tunica of the shoot apical meristem coordinate morphogenetic events. *Development*, 125, 1477–1485.

Scholander, P. F., Hammel, H. T., Bradstreet, E. D. and Hemmingsen, E. A. 1965. Sap pressure in vascular plants—negative hydrostatic pressure can be measured in plants. *Science*, 148, 339–346.

Singh, R. K., Svystun, T., Aldahmash, B., Jonsson, A. M. and Bhalerao, R. P. 2017. Photoperiod- and temperature-mediated control of phenology in trees—a molecular perspective. *New Phytologist*, 213, 511–524.

Taiz, L., Zeiger, E., Møller, I. M. and Murphy, A. 2022. *Plant Physiology and Development*, Seventh Edition. Oxford, Oxford University Press.

Tang, A. C. and Boyer, J. S. 2008. Xylem tension affects growth-induced water potential and daily elongation of maize leaves. *Journal of Experimental Botany*, 59, 753–764.

van der Schoot, C. and Rinne, P. L. H. 2011. Dormancy cycling at the shoot apical meristem: transitioning between self-organization and self-arrest. *Plant Science*, 180, 120–131.

Westgate, M. E. and Boyer, J. S. 1984. Transpiration-induced and growth-induced water potentials in maize. *Plant Physiology*, 74, 882–889.

Worrall, J. 1973. Seasonal, daily, and hourly growth of height and radius in Norway spruce. *Canadian Journal of Forest Research*, 3, 501–511.

CHAPTER 10

The ageing plant

10.1 Introduction

Trajectories of plant weight with age show a decline after an initial increase (Chapter 1), but the pattern of decline varies between species, with no single explanation for it applicable to all plants. This decline is associated with *ageing*—a general term that includes 'the time-based process of growth and differentiation as well as maturity, senescence and mortality' (Thomas 2013). *Senescence*, which has the general meaning of the process of growing old (according to the Oxford English Dictionary), has been used in reference to different processes, although much of the research has focussed on decline in fecundity and increase in the probability of death. Most research into senescence has been conducted with animals and Gaillard and Lemaître (2020) show that in plants senescence patterns for different characteristics may be asynchronous. In botany *senescence* is most frequently applied to the actively regulated loss from an individual of organs, particularly foliage, which is a controlled anatomical and physiological process (Lim et al. 2007).

Nuclear changes have been proposed as contributing to the ageing process. Medawar (1952) suggests somatic mutations could accumulate with age and become deleterious to an organism. Lanner and Connor (2001) examined trees of bristlecone pine in the White Mountains, California (Chapter 2) and Dixie National Forest, Utah with dated ages over the range 225–4,713 years but report no significant differences with age for metrics that could be affected by somatic mutations.

Chromosomes of eukaryotes have telomeres at their ends—long sequences of nucleotide repeats that protect the chromosome from deterioration or fusion with other DNA molecules. However, telomeres shorten with each DNA replication occurring in cell division and this shortening has been associated with the ageing process (e.g. Harley et al. 1992). An enzyme complex, telomerase, can add nucleotide repeats to the ends of telomeres and so compensate for the continual shortening. Flanary and Kletetschka (2005) found no age-related decline of telomere length in bristlecone pine and no decline in telomerase activity in needle and root samples but a decrease in cored samples. From arboretum samples they conclude that both telomere length and telomerase activity correlate positively with species expected lifespan and suggest 'an active role for telomere length regulation and telomerase activity in contributing to the longevity of trees', with bristlecone pine having longer lengths than other species and continued telomerase activity at old ages. While this may contribute to the species longevity, as they suggest, it does not explain the overall process.

Ageing of plants is associated with their reproduction, both in the distribution between asexual and sexual forms and in the material and energy involved. Asexual reproduction can occur through a number of morphological processes, such as tillering or shoot growth from creeping rhizomes, that produce new *ramets* (e.g. de Kroon et al. 2009); the original plant and all ramets together are referred to as a *genet*. Some genets can have extended existence over many years through this process. Generally, plants with clonal forms of growth have greater longevity than non-clonal plants (de Witte and Stöklin 2010).

In sexual reproduction a broad division can be made between semelparous plants (also

The Dynamics of Plant Growth. E. David Ford, Oxford University Press. © E. David Ford (2023). DOI: 10.1093/oso/9780192867179.003.0010

termed monocarpic)—most frequently annuals—that reproduce only once in their lifetime and iteroparous plants (also termed polycarpic)—typically perennials—that reproduce sexually multiple times over their life cycle. Both reproductive systems illustrate control of interactions between physiological processes and changes in anatomy, morphology, and development that occur as plants age.

Semelparous plants typically concentrate resources into their terminal reproductive effort and this is integrated with senescence of non-reproductive organs so that 'the plant dies in an explosion of monocarpic senescence' (Thomas 2013). As an example, the model plant *Arabidopsis thaliana* can complete its life cycle in six weeks and produce around 2,000–6,000 seeds per plant. Many grass species (Gramineae) occur as semelparous individuals, although many can also reproduce vegetatively. The example of control of grain yield of maize is considered here—grain filling occurs as day length and radiation decline and the development of the plant accommodates to these changes.

Iteroparous plants are typically perennials and have growth sustained over a substantial period. Examples are *Pinus longaeva* (Chapter 2) and *Thuja plicata* (Chapter 7), but many other tree species can live for hundreds of years and attain considerable size. Observations of increase in size have frequently led to questions about which growth processes may be most affected or that may limit size and so possibly longevity. A variety of patterns of age-correlated changes in mortality, reproduction, and physiological characteristics have been found. Most species in the 22 studies examined by Roach and Smith (2020) showed age-dependent decline in at least one trait, but they conclude that biological characteristics such as 'dynamic resource pools, the impact of size, dormancy and below-ground storage may influence patterns'. The processes involved in water movement through tree branching systems as they increase in size are discussed along with their effects on growth and the morphological and anatomical changes that may enable continued growth with increasing plant size.

10.2 *Zea mays* as a semelparous plant

Two important changes in breeding for increased yield in maize (Chapter 2) were: (i) decrease in anthesis to silking interval, which combined with a reduction in tassel size has been interpreted as affecting the numbers of kernels set; and (ii) increase in the staygreen character, which implies extension of the period of photosynthesis but which also has implications for remobilization of material in leaves, particularly N, that could affect grain yield.

A point of recurring interest of practical importance is whether yield of maize is source limited (controlled by foliage amount and rate of photosynthesis) or sink limited (controlled by the demand of the number of grains set). Although there can be differences in seed dry weight between crops, the number of ears per unit area of land is the dominant component of yield determination (Borrás et al. 2004).

From the time of grain set the ear becomes the major sink for carbon (C) and nitrogen (N) compounds. Moss (1962) found that preventing pollination by bagging ear shoots and silks reduced photosynthesis to only 55% that of normal plants after one month and that sap from the stalks of barren plants had 162% of the sugar content relative to that from normal fertile plants, suggesting that the reduction in rate of photosynthesis is due to end-product inhibition. Foliage on barren plants had delayed leaf senescence.

10.2.1 Establishment of the ear as a sink, kernel set, development, and growth

Following pollination there is a period of some 12–15 days of active cell division and differentiation and DNA endoreduplication within the kernel and formation of endosperm tissue but with little or no increase in weight. Source availability per kernel at the time of kernel number determination defines maize potential sink capacity and final kernel weight (Gambín et al. 2006; Borrás and Gambín 2010). Gambín et al. (2006) used a calculation of whole plant growth rate per kernel as an indicator of resources available during kernel formation. An examination of 12 genotypes, each at two sites and under full irrigation, showed that final kernel weight was correlated with plant growth rate per kernel at flowering.

Evidence for the importance of sink limitation in determining weight gain during grain filling comes from experiments manipulating the plant by reducing, or shading, foliage amount (source limitation) or reducing ear size (sink limitation). Borrás et al. (2004) summarized the results from 16 such investigations: when source strength was reduced, seed weight decreased. However, when assimilate availability per seed was increased during the post-flowering period, seed weight only increased marginally. See also Ordóñez et al. (2018).

Sink control during the grain filling period can occur through control of unloading of sucrose from the phloem at the ear that in turn can affect sucrose loading into the phloem at the leaf. Taiz et al. (2022; Chapter 12) review loading into the phloem, transport, and unloading. Borrás and Gambín (2010) summarize the process of sucrose unloading from the phloem and transport of sugars to the developing endosperm of the kernel as first requiring symplastic unloading and transport from phloem sieve elements to efflux cells in the pedicel. There are no plasmodesma connections between pedicel cells, on which the seed sits, and the endosperm of the kernel, so transport to the influx cells of the kernel is apoplastic. The efflux of sucrose is driven through sucrose hydrolysis by an apoplastic invertase enzyme to produce hexoses (Lemoine et al. 2013). Borrás and Gambín suggest: 'This insoluble acid invertase plays a key role in hydrolysing sucrose unloaded at the terminal sieve tube element and is thought to be essential for establishing sink strength via the maintenance of a favourable concentration gradient between the sieve tube and the apoplast.' Whilst kernel growth requires sugars for filling the endosperm, largely with starch, there is a limit to how much sugar can be used. Uptake of sugars by kernels is intimately related to kernel water content and Borrás and Gambín note that the ability to regulate water and sugar contents by growing kernels ensures stability in the process of endosperm development.

10.2.2 Nitrogen uptake and remobilization

During the vegetative phase of growth N is stored within the plant and then subsequently transferred to the grain. Leaf N is an important component of total N in the plant and more recent staygreen hybrids have a changed N economy.

A conceptual model of patterns of N occurrence within a developing maize plant, based on a hybrid used in the 1970s, defines different parts of the plant as sinks for N during vegetative growth and then becoming sources for the grain, i.e. involving remobilization (Fig. 10.1).

Crawford et al. (1982) estimated fluxes between component parts of the plant from pollination onward to 12, 24, and 36 days after pollination during which time the ear was filling. In their investigation all plants received NO_3^- during vegetative growth: from pollination one group of plants received no N and the other 3.75 millimolar ^{15}N as NO_3^- (called exogenous N). The role of N previously accumulated in different parts of the plant (called endogenous N) could be distinguished between those plants receiving N from the time of pollination and those not. The husk, cob, and shank first accumulated N, so acting as sinks, and then became sources of N for the grain, and this suggest a function of these plant parts is 'to accumulate N in the early stages of reproductive growth when total N flux exceeds N flux into the grain'. As total N flux into the grain increases, husks, cob, and shank act as sources of N.

Interestingly, both groups of plants, those with and without N in the root solution after pollination, accumulated the same grain dry matter and N in the grain. The primary source of endogenous N was the stalk, with a faster initial than subsequent rate of loss, followed later in growth by loss of N from lower and upper leaves. Crawford et al. suggest that this sequence indicates evolution to delay net loss of N from photosynthetic organs, enabling a longer period of photosynthesis, whether there is a root supply of N or not. However, N in the rooting medium decreased rates of loss of both exogenous and endogenous N from the stalk and lower and upper foliage.

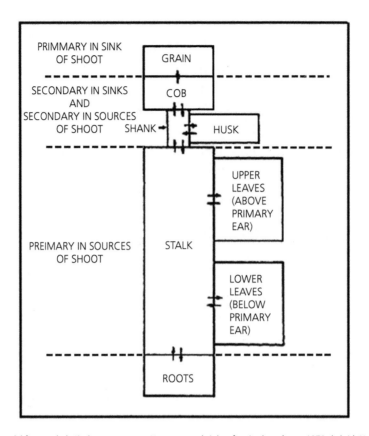

Fig. 10.1 Conceptual model for morphological components as N sources and sinks of maize based on a 1970s hybrid. N accumulates in the stalk prior to pollination but is then translocated to secondary sinks close to the grain that become sources for the grain. More N is translocated when no N is available from roots. Loss of N accumulated in leaves prior to pollination, i.e. a primary source, increases as time progresses from pollination; plants with no N supplied to roots after pollination lose N from leaves more rapidly. The husk, shank, and cob, on which grains are mounted, accumulate N from the time of pollination. The shank acts as a sink for N during the first phase of reproductive growth, ~24 days. The husk, shank, and cob are sources of N for the expanding grain. Reproduced from Crawford et al. (1982) with permission from Oxford University Press.

With a 1980s hybrid Pan et al. (1986) found that plants with two ears, and so likely greater sink strength, had greater grain production than plants with one ear, as well as greater N remobilization from stalk and roots and N translocated to grain but that remobilization from the leaf was not affected. Cliquet et al. (1990) confirm the role of stalks in N storage and remobilization described by Crawford et al. but conclude that the same processes do not apply to C accumulation in the kernel which they suggest comes from photoassimilates accumulated after anthesis.

The contributions of remobilized and soil sources of N to the grain have changed as hybrids have been developed since the 1990s (Ciampitti and Vyn 2014). Leaves are a source of N within the plant and leaf senescence is involved both in remobilization of N and reduction in the rate of photosynthesis. Rajcan and Tollenaar (1999a, 1999b) conclude: 'increased leaf longevity of a new relative to an old hybrid was associated with a larger source:sink ratio during grain filling'. They compared dry matter accumulation, its distribution between plant parts, and N metabolism during grain filling between two hybrids: one released for commercial planting in 1959 (old) and the other in 1988 (new). Plants were exposed to different levels of N fertilizer and manipulations to reduce the source of photosynthates (partial removal of upper ears) or sink (bagging of female flowers and partial ear removal). Total N uptake of the new hybrid was 10–18% greater—attributed to post-silking N uptake being 60% for new compared to 40% for the old hybrid.

In a comparison between the same two hybrids used by Rajcan and Tollenaar, Ma and Dwyer (1998) found the newer hybrid, which they referred to as a 'staygreen', had a greater nitrogen use efficiency (NUE) broadly defined as N in plant grain yield relative to the amount of N applied (Oenema 2015), with 24% more N uptake during grain filling. They suggest the new hybrid 'maintained photosynthetic production longer by maintaining leaf N concentration, which in turn produced more dry matter for grain filling and diluted the grain N concentration'. Kosgey et al. (2013) found a trend of greater remobilization of N from leaves at physiological maturity of lower staygreen rated hybrids combined with greater N accumulation in kernels.

However, delayed foliage senescence does not necessarily improve post-silking C accumulation. Acciaresi et al. (2014) attribute lack of gain in yield in maize hybrids with differences in staygreen used in Argentina post-2000 due to the senescence differences occurring at low canopy levels where, because of low light levels, greater leaf N does not produce increased C fixation, and suggest that to obtain increased yield such attributes need to be combined with vertical leaf characteristics.

Using ^{15}N application to a staygreen hybrid of maize, combined with source and sink manipulations, Yang et al. (2016) found pre-silking N contributed at an early grain filling stage and the post-silking N uptake was largely regulated by post-silking dry matter accumulation in both grain and vegetative organs. Remobilization of pre-silking N was regulated by grain N demand, but the remobilization from leaves was not dependent upon the existence of grain.

Borrell et al. (2001) summarize the effect of the staygreen character and implicate the importance of grain set in the establishment of the sink:

We hypothesise that increased N uptake by stay-green hybrids is a result of greater biomass accumulation during grain filling in response to increased sink demand (higher grain numbers) which, in turn, is the result of increased

radiation use efficiency and transpiration efficiency due to higher SLN (specific leaf N).

They suggest delayed leaf senescence results in more C and N reaching roots of staygreen hybrids during grain filling, so maintaining a greater capacity to obtain N from the soil compared with senescent hybrids.

In maize, morphological development and N and C economy are integrated in the process of seed production—the essential process for a semelparous plant; this is under genetic control and there is not a unique organization for seed production. Seed set is the principal driver of the system as it establishes sink strength, but the sources of C and N during grain filling can vary and potentially affect yields. The dynamics of the process are centred on seed production. The longevity of the parent plant is a consequence of this process.

10.3 Iteroparous plants

Trees are iteroparous species widely studied in relation to the effects of ageing on their growth. Timber yield of forest plantations reaches a peak early in the life of a stand and then gradually declines, which typically is described in the reduction in number of trees due to thinning or competition mortality (Assmann 1970). However, growth rate of individual trees also declines and botanical explanations for this decline have also been sought. A frequently expressed view, at least in conversation, is that trees die because of decline in their ability to produce sufficient resources to sustain their existing body. This is a simplistic view not supported by research.

Three lines of investigation are considered:

- demand for resources increases beyond the capacity of the plant to satisfy it;
- a decline in photosynthetic rate; and
- increase in water stress of growing tissue.

Rather than consider how iteroparous plants may die an alternative is presented here that, as environmental changes occur with increase in size, growth is maintained through modifications in anatomy and morphology of tissue and branches produced.

10.3.1 'Death-by-starvation'

For most trees an increase in age brings an increase in size. Death related to this increase has been referred to as 'death-by-starvation' (Molisch 1938; Rosbakh and Poschlod 2018) which Rosbakh and Poschlod summarize as: 'a plant dies due to exhaustion or starvation of its source organs (e.g. photosynthetic leaves and stems) caused by the demands of sink tissue, such as flowers, developing seeds, tubers and other structures'. This suggests that maintaining existing growth comes to require so much resource that production of new foliage declines, with a corollary that reduction in growth will extend lifespan by curtailing the rate at which 'exhaustion or starvation' is reached—an explanation proposed for such observations made of greater longevity in some stressful environments (Rosbakh and Poschlod 2018).

Ryan et al. (1997) examined this theory in relation to C: it was supposed that as the amount of wood increases the respiration required to maintain it alive (R_m, maintenance respiration) increases so that primary net primary production decreases and there is less C available for growth. Proposals were made that R_m depends upon wood mass or cambial area, which increases with age, and also that there may be some decline in foliage amount that would also reduce gross primary production (GPP).

Using temperature as the primary driving variable and sapwood as the respiring tissue, so discounting the central heartwood, Ryan and Waring (1992) estimated stem and branch R_m and compared it to stem and branch growth for even-aged stands of *Pinus contorta* (lodgepole pine) of 40, 65, and 245 years growing closely together at 2,800 m elevation in Colorado, the United States. Estimates of stem and branch growth and their associated construction respiration decreased markedly: 210, 124, and 46 g m^{-2} year^{-1} for stem over the age series, while branch respiration increased only slightly from 61 to 79 g m^{-2} year^{-1} from 40 to 245 years. Ryan and Waring conclude that these differences in respiration are insufficient to explain the decline in wood production. Foliage amounts were estimated, respectively, as 12.3, 12.1, and 7.7 m^2 m^{-2} (all sides of needles) and calculated rates of wood growth declined, whether calculated on a foliage area or light absorption basis. Ryan and Waring suggest the decline was due to some other factor than decline in foliage area. They consider an increase in root growth may be one contributing cause but that a reduction in rate of photosynthesis is more likely.

Ryan et al. (2004) followed this with an investigation of plantations of *Eucalyptus saligna* at Hilo, Hawaii, where it grows rapidly and can achieve a height of ~22–24 m and an L of 8.2 in 6 years when grown at close spacing (10,000 trees ha^{-1}) and with intensive fertilization. Peak above ground production occurred at 1–2 years and declined from 1.4 at age 2 to 0.60 kg C m^{-2} year^{-1} at age 6 years. Above-ground wood respiration (R_m) was not sufficient to account for decline in above ground production. It declined as also did total below C (growth plus respiration). The decline in above ground wood production by year 6 (42% of peak value) was greater than the decline in photosynthetic capacity (64% of peak). Both below ground C and foliage respiration increased. Ryan et al. suggest decline was not caused by nutrient limitation, decline in L, or photosynthetic capacity and that a number of factors could lower production, including stomatal closure related to tree height, or other limits to photosynthesis, or a change in branching patterns or leaf demography.

10.3.2 Decline in photosynthesis

Ryan et al. estimated a decline of 15% in photosynthetic rates per unit leaf area could account for the reduction in the above ground growth found in lodgepole pine stands (Ryan and Waring 1992). Net photosynthesis per unit foliage area of one-year sunlit foliage from the mid-canopy averaged 14–30% lower on foliage from older trees in the 45- and 275-year stands of lodgepole pine and in pairs of *Pinus ponderosa* aged 54 and 229 years (Yoder et al. 1994). Yoder et al. conclude that the lower net photosynthetic rate observed in older trees could explain lower wood production but were careful to note that causality could be that 'growth reductions due to any other reason could lower net photosynthesis due to a smaller carbon sink'.

Stomatal conductance in these trees showed similar reductions to that of net photosynthesis and differences in both were most pronounced in mid-morning. In general, larger average differences were found between young and older trees when average net photosynthesis and stomatal conductance were lower. Mean C isotope ratios, ^{13}C and ^{12}C, referred to as δ^{13}C, for one-year foliage from young and old trees were −27.2 and −26.0. Greater values indicate relatively less discrimination between isotopes by the carboxylating enzymes which is attributed to greater stomatal closure and so longer residence of gas molecules within the air spaces of the leaf. Mean δ^{13}C for wood from young and old trees was −25.8 and −24.5 and values increased (became less negative) with height on the tree. Yoder et al. note that stomata of older trees began closing about two hours earlier than those on younger trees but that leaf water potentials were nearly identical at this time and conclude that hydraulic conductance of the vascular system was lower for older trees and this was the likely cause for observed differences in stomatal conductance and net photosynthesis.

Subsequently, Hubbard et al. (1999) found mean whole tree sap flow per unit leaf area to be 53% lower in old trees in the ponderosa pine stand and calculated mean hydraulic conductance to be 63% lower. Similarly, Drake et al. (2010) conclude that decline in photosynthetic rates and stomatal conductance in loblolly pine (*Pinus taeda*) over a stand age range of 14–115 years was related to increase in hydraulic limitation and not N availability that could affect photosynthetic capacity.

In a review of age-related changes in net photosynthesis of trees and shrubs Bond (2000) reports 13 published studies showing a decrease in net photosynthesis with increased age and 7 showing an increase but noted that 4 of those 7 may be due to an increase in available water as roots reached deeper water sources. Niinemets (2002) composed a data set of published photosynthesis measurements: 125 for *Picea abies* and 80 for *Pinus sylvestris*. Net assimilation rates were strongly and negatively related to height of tree. Stomatal conductance also declined with increase in height, but so too did maximum carboxylase activity of RuBisCo.

In contrast to conifers, broadleaved species generally have higher foliage area-based photosynthetic rates in larger trees, although leaf mass-based values may be greatest in sapling-sized trees (Sendall and Reich 2013); while many studies have shown changes with increasing tree size in leaf characteristics consistent with declines in leaf function not all species show the same pattern and decline in stomatal conductance with height is not always sufficient to explain reduction in photosynthesis.

Leaf photosynthesis (measured at photon value 1,500 μmol m^{-2} s^{-1}) varied with height in a non-linear way on a log photosynthetic rate:log height basis for red maple (*Acer rubrum*), a shade-tolerant species, northern pin oak (*Quercus ellipsoidalis*), mid-tolerant, and black cherry (*Prunus serotina*), less tolerant, in a mixed deciduous forest. Contrary to Sendall and Reich's initial expectations of a decline with height, there were generally curvilinear relationships, although they did not conclude there was a consistent maximum at mid-height of the tree as reported by Thomas (2010) for each deciduous species he measured in a mixed deciduous selection forest in Ontario, Canada. Sendall and Reich (2013) explicitly exclude a unifying theory to explain the variation in properties they found based on the measurements they made.

Sala and Hoch (2009) suggest that the proposed hydraulic limitation of photosynthesis to explain reduction of height growth through reduced photosynthesis implies there should be reduced concentrations of non-structural carbohydrates (NSC) with increase in height of trees. They measured NSC in leaves, branches, and boles of ponderosa pine of different heights and at two sites and found height-related growth reductions associated with significant increases in NSC and greater increase on drier relative to a moist site. Periods of reduced or zero growth generally show greater NSC and Körner (2003) concludes that generally tree growth does not seem C limited. Water deficits generally increase C concentrations in plants and Muller et al. (2011) suggest this could be because organ expansion is reduced at less negative ψ than photosynthesis, although there can be differences between species in these relationships (Fig. 10.2). Muller et al. suggest there can be a transition from source to sink control under water deficits—so a decline in rate of photosynthesis can be the result of a decline in growth.

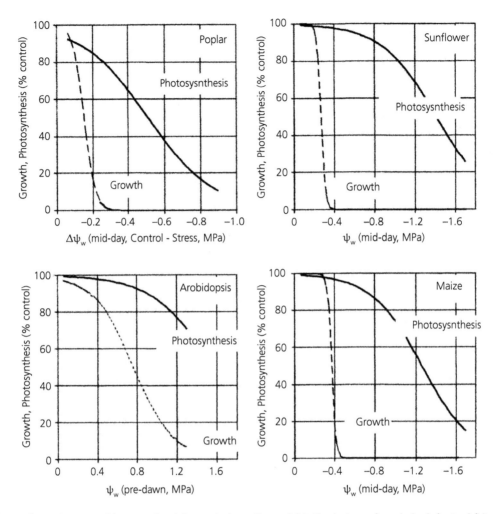

Fig. 10.2 Differential sensitivity of shoot growth and photosynthesis to soil water deficit. The abscissae refer to the level of water deficit expressed as mid-day leaf water potential difference between stressed and non-stressed plants (poplar), mid-day leaf water potential (maize, sunflower), and pre-dawn water potential of the whole rosette (*Arabidopsis*). The ordinate axes refer to both photosynthesis (plain lines) and organ growth (dashed lines), both as a percentage of a well-watered control. Abscissae and ordinate axes do not always refer to the same variables; growth is expressed as shoot height increase in poplar, rosette size at bolting in *Arabidopsis*, and leaf elongation rate during day periods in maize and sunflower. Reproduced from collated data presented by Muller et al. (2011) with permission from Oxford University Press.

10.3.3 Water potential effects on growth

The failure of explanations for growth decline based on C economy alone has led to an examination of direct effects of decline in water potential on growth. The diurnal progression of water potential in trees 7 m tall, how it may vary if soil water availability declines, and effects on shoot extension are illustrated in Chapter 9. As trees increase in height, resistance to water movement increases,

lower water potentials may occur, so that water in the xylem is under tension, and there can be increase in the possibility of cavitation—where a conducting cell loses water and becomes filled with air. Under such conditions there is the possibility that cavitation can spread from cell to cell resulting in an embolism that would reduce conductivity through the tissue. Stomatal closure can reduce the development of low water potentials and obviously may have an effect on reducing photosynthesis, but, for

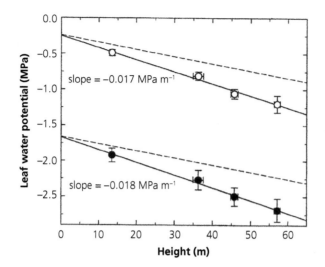

Fig. 10.3 Pre-dawn (open symbols) and mid-day (filled symbols) foliage water potentials with height above ground on combined measurements from young and old *Pseudotsuga menziesii* in July 2001. Bars represent standard errors for measurements on three samples at each location from each of five trees (*n* = 15). Dashed lines represent the theoretical gravitational gradient in the absence of transpiration (−0.01 MPa m⁻¹). Reproduced from Woodruff et al. (2004) with permission from Wiley.

growth, that reduction may be of secondary importance relative to the direct effect on cell expansion.

In old-growth Douglas fir, leaf water potential (ψ_l) has been found to decrease with increasing height on the tree (Fig. 10.3). Woodruff et al. (2004) determined the amount of water in tissues in relation to ψ_l by slow dehydration while making repeated determinations of ψ_l. Osmotic potential of the expressed sap was measured and from these measurements turgor estimated (for details see Woodruff et al.). Turgor decreased with increasing height particularly during swelling of buds. Vertical changes in branch elongation, leaf dimensions, and leaf mass per unit area (LMA) were consistent with turgor limitations with increasing height. Although osmotic increase with increase in height developed as the season progressed, it was not sufficient to compensate for the vertical gradient of ψ_l (see also Meinzer et al. 2008 for needle elongation).

Similarly, in coastal redwood (*Sequoia sempervirens*) both pre-dawn and mid-day turgor declined with increasing height over a range of ~50 m to ~112 m (Koch et al. 2004). A marked change in shoot and foliage morphology was found with increasing height (Fig. 10.4) along with an increase in LMA. Koch et al. found in a multiple regression analysis LMA to be correlated with decline in

pre-dawn water potential but only marginally with an estimate of light conditions, and Ishii et al. (2008) suggest hydraulic constraints on morphological development affects light utilization in photosynthesis in these trees.

10.4 Changes in plant structure and development with age

Two aspects of structure can change as growth takes place and alter the dynamics of plant water relations or the effects they may have on longevity: wood anatomical structure in the trunk that can affect the potential for embolisms as well as water conductivity; and branching structure, particularly in reiteration of branches.

10.4.1 Wood structure

Three processes involved in water movement vary with tree height and can be affected by wood structure:

1. conductance that varies with tissue water content;
2. capacity for water storage, its movement into the transpiration stream, and its recharge; and

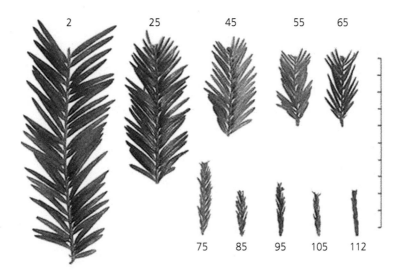

Fig. 10.4 Variation in leaf structure with height in *Sequoia sempirvirens*. Leaf length and the angle between the long axis of the leaf and supporting stem segment both decrease with height. Numbers denote sample height in metres. Scale divisions are in centimetres. Reproduced from Koch et al. (2004) with permission from *Nature*.

3. embolisms, i.e. cells that fill with gas when water potential decreases and so tension increases, that may occur in groups and can develop with a decrease in tissue water potential and reduce conductance but cells may subsequently recharge with water so the embolism is removed.

Hydraulic conductivity (k_s) is not constant for a piece of wood but declines curvilinearly as negative pressure is applied. The metric most used in describing this is percentage loss of conductivity (PLC) where the baseline is conductivity for saturated wood. A 12% loss in conductivity is considered the standard value at which air enters and initiates embolism (Domec and Gartner 2002). PLC varies both with wood structure, and consequently with height on the tree, and with times of day and year as environmental conditions change.

Domec and Gartner calculated variation in PLC as negative pressure was applied on wood from Douglas fir. Calculations for winter and pre-dawn summer PLC have smaller values as mid-day summer at nodes 5, 15, and 35 (the base of the live crown) from the top of the tree and as well as at the base of the tree (Fig. 10.5) (Domec and Gartner 2002). However, the values towards the top of the tree show less reduction in PLC than those at the base of the

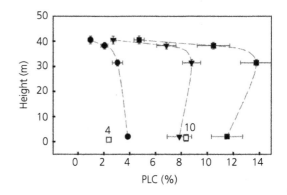

Fig. 10.5 PLC height in *Pseudotsuga menziesii* trees in relation to height. Symbols: closed circles, winter; inverted triangles, summer-predawn; and closed squares, summer mid-day values; respectively, for trunks of mature trees, and summer-midday values for saplings and seedlings. Numbers 4 and 10 alongside open squares denote 4-year-old seedlings and 10-year-old saplings, respectively. Error bars = ±1 SE. Reproduced from Domec and Gartner (2002) with permission from Oxford University Press.

live crown and base of the trunk, both of which were more vulnerable to cavitation, i.e. closer to a value of PLC of 12% at mid-day. They attribute the lower PLC values at the tree top and within the tree crown to a greater proportion of juvenile wood which is produced near to the pith, approximately the first 20 annual growth rings, with mature wood being ~≥35 rings from the pith, and k_s of juvenile wood was 11% lower but resistance to cavitation was greater.

Domec and Gartner (2002) compared the effects of Douglas fir wood structure on both hydraulic safety—PLC values that would result in embolisms—and mechanical safety—the potential for buckling and height limitation—and concluded that wood structure in Douglas fir is a result of selection for hydraulic rather than mechanical safety.

There are differences between species in hydraulic conductivity (e.g. McCulloh et al. 2019) and aspects of wood structure of ageing trees that can affect water relations. Domec and Gartner (2003) made similar measurements and calculations on ponderosa pine trees to those they had made on Douglas fir, with trees aged 30 (young), 70 (mature), and 220 (old) years. This species generally has large sapwood and they found outer sapwood generally had 25–60% greater k_s than inner sapwood and the water potential at which embolism started was 1–3 MPa less. Capacitance increased with an increase in k_s and in vulnerability to embolism and k_s at the point just below the live crown increased over the age sequence: respectively 1.4, 5.1, and 8.0×10^{-12} m^2.

Explanations for variation in k_s with height have been sought considering two anatomical properties: diameter of conducting cells in the xylem that can change with height and structure and distribution of pits in cell walls through which water movement between cells takes place.

Eucalyptus regnans in south-eastern mainland Australia and Tasmania is the hardwood species with the largest height of up to 100 m (Petit et al. 2010). Petit et al. found that vessel area in sapwood adjacent to the cambium, D_h mean hydraulic diameter, increased with distance from the apex of the tree while the number of vessels per unit sapwood area decreased. D_h increased six- to ten-fold from the apex towards the point of crown insertion, i.e. the base of the live crown, but then increased more slowly. Petit et al. estimated that the sharply tapered xylem at the apex of the trees accounted for the major amount of whole tree resistance and report that the rate of change in D_h from the apex of *E. regnans* is greater than that reported for species of similar heights. Water in the xylem at the top of the trees is under considerable tension which would

limit the formation of wide cells that may increase risk of xylem failure by cavitation. They consider that capacity for xylem acclimation is an evolutionary adaptation for the tall growth of this species on well-structured soils in a region with considerable monthly rainfall.

Water conduction in the xylem of gymnosperms is through tracheids which, unlike the vessels of angiosperms, do not have perforated endplates and movement of water between tracheids occurs through co-located pits through the walls of neighbouring cells. Pits, which are broadly circular, have a torus–margo membrane: a porous outer net-like margo and a central impermeable torus. The membrane can be deflected so that the torus blocks the aperture of the pit and, at full deflection, this can prevent flow between cells. If one cell has become air-filled, and so is at near atmospheric pressure, but the neighbouring cell is water-filled and at negative pressure the margo will stretch and the torus will seal the pit aperture, preventing an embolism from spreading.

There is variation between species, environments, and within trees, of cell dimensions and pit frequency and dimensions. For example, Aumann and Ford (2006) summarize published differences between Douglas fir growing in coastal and interior (drier environment), including tracheid lengths respectively 4.0–5.9 and 3.0–3.6 mm, early wood tracheid widths 40–56 and 36–41 μm, and number of bordered pits per tracheid ~144 and ~65.

Domec et al. (2008) investigated bordered pits in the trunk and branches of Douglas fir ranging in height from 6 to 85.5 m. Pit torus diameter changed little with increasing height, but pit aperture diameter decreased significantly so that the ratio of torus diameter:aperture increased. This ratio is a measure of torus overlap at aspiration and where there is larger overlap there is greater resistance to embolism, but this occurs with a reduction in water transport efficiency. Domec et al. estimated this to reduce by a factor of 3 over a height gradient of ~80 m.

10.4.2 Branching structure

The flow of water through the whole organism is controlled by differences in leaf-specific

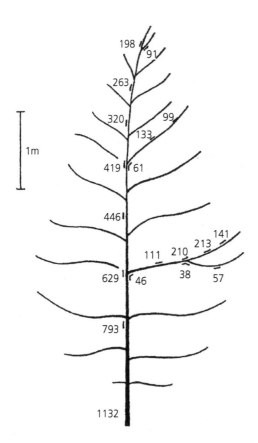

Fig. 10.6 Pattern of LSC in the crown of a *Tsuga canadensis* in 10^{-6} kg s^{-1} m^{-1} MPa^{-1}. LSC values are higher in the trunk than in branches and higher in first-order than in second-order branches. Branch junctions have hydraulic constrictions. Reproduced from Tyree and Ewers (1991) with permission from Wiley.

conductivity (LSC) of branch segments, i.e. k_s divided by the area of foliage distal to the segment (Tyree and Ewers 1991). LSC values are greater for trunks relative to branches and decrease with increasing height along a trunk (Fig. 10.6). Tyree and Ewers note that in each of three species of gymnosperms, *Tsuga canadensis*, *Abies balsamea*, and *Thuja occidentalis*, LSC increased with larger stem or bole diameter: LSCs of ~0.3 × 10^{-5} kg s^{-1} m^{-1} MPa^{-1} for 1-mm-diameter stems and ~30 × 10^{-5} kg s^{-1} m^{-1} MPa^{-1} for 300-mm boles and generally smaller-diameter stems have smaller-diameter tracheids than larger-diameter stems. This difference in LSC requires that a pressure gradient needed to maintain water flux to transpiring foliage distal to the smallest stem segments needs to be 30–300 times

steeper than the corresponding gradients at the base of boles. The effect of this pattern is that most of the resistance to water flow is in the last metre, or less, of pathway from the base of the tree, and Tyree and Ewers comment: 'Consequently, most small branches can compete for water on a more-or-less equal basis, i.e. each small branch behaves like an independent plant rooted in a common, highly conductive bole.'

This raises the question of the extent to which branches and/or foliated branchlets have autonomous water conduction systems. Where there is such autonomy, branches or connected groups of foliage may experience different degrees of water stress and possibly differences in survival. Separation of water-conducting pathways is referred to as segmented or sectored. Tyree and Ewers (1991) define Zimmermann's (1983) view of segmentation as 'any structural feature of a plant that confines cavitations to small, distal, expendable organs in favour of larger organs representing years of growth and carbohydrate investment'. However, structural differences can have a more comprehensive effect than just variation between central and more peripheral parts of the tree. Larson et al. (1993, 1994) describe separation of water conducting pathways extending all the way from root to foliage and refer to this structure as radially sectored xylem pathways.

Hydraulic systems within branches of Douglas fir are hydraulically relatively unconnected. Brooks et al. (2003) reduced transpiration on between 85 and 90% of foliage by shading on each of 10-, 20-, and 45-year branches and measured g_s and photosynthesis during days with typical late summer transpiration load. Shading to reduce the amount of transpiring foliage did not increase g_s of the remaining sunlit foliage, and similarly photosynthetic rates were unchanged in six of the seven trees examined. From measurements of hydraulic conductivity of branch segments after the physiological investigations were complete, flow through lateral branches increased by 15% rather than the expected 50% if the hydraulic systems had been completely interconnected. Schulte and Brooks (2003), using dye injection techniques, also conclude that there is limited hydraulic interconnection within branches of Douglas fir and ponderosa pine and discuss variation in the results of this type of experiment conducted with different species.

10.4.3 The role of morphology in sustaining growth

Trees have two forms of branching: sylleptic, produced by a continuing axis and its buds; and proleptic, produced by delayed growth—as in the long shoots that develop from short shoots of bristlecone pine and ensure continued foliage production. Epicormic buds are frequently a source of proleptic branching and occur in many angiosperm and gymnosperm species (Meier et al. 2012).

Development of epicormic buds into foliated branches has frequently been described in response to damage through wind, frost, or fire (Meier et al. 2012). However, this development also occurs as a continuing component of crown development without any apparent external stimulus and the production of foliated branches by this process is termed adaptive reiteration (Bégin and Filion 1999; Ishii and Ford 2001). Ishii and Ford report that in ~400-year-old trees of Douglas fir upper-crown branches

were still extending sylleptically while mid- and lower-crown branches had reached maximum lateral extension and were being maintained by adaptive reiteration along established axes. Sylleptic branching of Douglas fir from a main branch axis produces a frond of foliage (Fig. 10.7a) with a main axis from which laterals are produced, similar to that produced by Sitka spruce (Chapter 7; Fig. 7.8). The number of laterals produced at each annual increment depends upon the environment. Figure 10.7a illustrates a structure typical of mid-crown of an old-growth tree on which the main axis produces a new shoot extending the main axis and two laterals, one on each side of the main branch axis; these continue growth with a shoot extending the lateral axis. At some age laterals no longer producing extending axes, and some older foliage on laterals, and the main axis, die. Foliage on the frond is then separated from other foliage by non-foliated branch segments; Ishii and Ford term this structure a shoot cluster unit (SCU).

Epicormic buds are produced on the main axis at branching nodes and at the time of lateral bud formation, but these remain dormant when the laterals

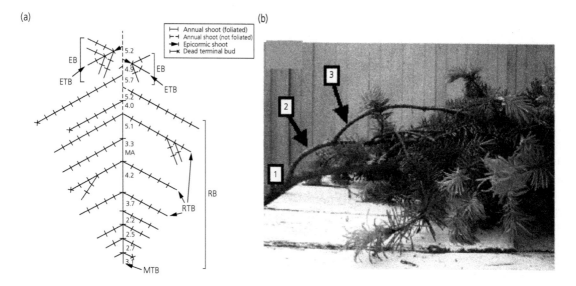

Fig. 10.7 (a) A map of a typical shoot cluster unit (SCU) of *Pseudotsuga menziesii* shows the main axis (MA), regular lateral branchlets (RB), and epicormics branchlets (EB) with their respective terminal buds (MTB, RTB, and ETB). Values along the main axis denote the distance between branchlets (in centimetres). Branchlets without a live terminal bud are marked with an X at the tip. (b) Accumulation of generations of SCUs. Black arrows point to a succession of new epicormic axes proliferating on top of the major axis of the parent, with generations assigned assuming the parent SCU is generation 1 (numbers in boxes). Part (a) reproduced from Ishii and Ford (2001) with permission from Canadian Science Publishing. (b) Photograph Dr. M. C. Kennedy.

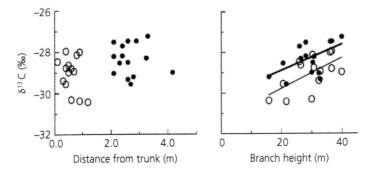

Fig. 10.8 Carbon isotope ratios ($\delta^{13}C$, ‰) of current-year needles on epicormics shoots (open circles) and on shoots at the distal end of the parent branch (closed circles) in 450-year-old *Pseudotsuga menziesii* trees. Where possible, pairs of shoots (epicormic and parent) were sampled from the same branch. Sample branches were chosen from the lower crown to minimize differences in light environment between shoots (paired *t* test of canopy openness: $t = 0.786$, $p = 0.447$). (a) Epicormic shoots are closer to the trunk than parent shoots and $\delta^{13}C$ values of current-year needles on epicormic shoots were significantly lower than those of current-year needles on the parent ($t = 3.610$, $p = 0.003$), indicating lower time-averaged water-use efficiency (carbon gain per water transpired), i.e. less water stress for epicormic shoots. (b) $\delta^{13}C$ values decreased with increasing height. Similar height-related changes in $\delta^{13}C$ were found by Winner et al. (2004). A linear mixed effects model (*lme*, R statistical software) with random slopes grouped by branch indicated that $\delta^{13}C$ values increased with increasing branch height and were lower for epicormic shoots than for parent shoots ($p < 0.05$). Reproduced from Ishii et al. (2007) with permission from Oxford University Press.

grow in the following spring. Epicormic buds are invariably on the upper (dorsal) surface of the SCU. They remain dormant typically until the lateral shoots from the associated whorl no longer produce a lateral extension, although epicormic extension does not always follow that condition. Epicormic buds produce shoots above the plane of the SCU from which they grow and typically develop an axis upwards at a slight angle to their parent axis and their growth is initially less than that of shoots developing along the main axis. The number of years an epicormic bud remains dormant does not vary with position of an SCU within the crown. Typically the parent SCU continues to decline, producing smaller main axis increments and continuing foliage mortality, and epicormic shoots can develop into a new SCU. An effect of adaptive reiteration is that new foliage produced is closer to the main trunk of the tree than foliage towards the distal end of its parent SCU. This has an effect on tissue water relations, with the foliage of the new epicormic sprout having less water stress (Fig. 10.8).

Ishii and Ford (2001) found that on lower crown branches shoots were predominately produced through epicormic adaptive reiteration. In addition to reiteration of SCUs epicormic shoot production was found from large-diameter branches and the main tree trunk and produced branches

that can be distinguished by their irregular angles of connection to parent branches. These epicormic shoots form branches that develop as an inner layer to the first formed crown. The process of adaptive reiteration is likely to maintain the established crown of older Douglas fir after height growth and lateral crown expansion have stopped. The maximum number of successive generations of SCUs, g*, found in Douglas fir was 7 where a new generation sprouted from an SCU previously formed by adaptive reiteration and this may set a limit to the effectiveness of adaptive reiteration in sustaining growth. Using a simulation model for branch growth Kennedy et al. (2010) calculated g* under different growth constraints for branches and their foliage and they found a median g* of 8.9 that was driven by hydraulic restriction at epicormic junctions relative to restrictions formed by regular sequential growth.

10.5 Conclusion

After finding no cellular indication of senescence in bristlecone pine Lanner and Connor (2001) ask:

'What then of death? Is there no limit to the life of a non-senescing pine?
Must it not die someday, and if so of what?'

Some organisms, particularly animals, have clearly definable end-points to their lives. Humans typically have a stated cause of death on a certificate— usually some illness or disease—although there may be a number of underlying components to that cause. Some plants do die from pathogens, but what is of general interest is the decline in growth that most usually precedes death, frequently over a long proportion of a plant's existence. A plant's life is defined by continued meristematic activity and rather than asking how plants die it is more appropriate to ask how that activity is continued but may decline.

Clearly in semelparous plants such as maize there is a distinct developmental change from vegetative growth to sexual reproduction: the plant's structure and physiology function together to produce seed. Senescence and chemical changes, where they occur, are an integral part of seed production; senescence and death of the parent are consequences. Between plants with apparently similar patterns of growth there can be considerable variation in growth control. Borrás et al. show that yield is usually more sink than source limited during seed filling in wheat (*Triticum aestivum*), maize (*Zea mays*), and soya bean (*Glycine max*), but there are differences between these species that affect the roles of source and sink limitations in controlling seed growth. For example, the number of cells in the endosperm or cotyledons determined before or just after onset of rapid seed growth correlates well with seed size at maturity. Wheat and maize have maximum seed volume early in their development when they are ~35% of final dry weight, but soya bean seeds reach maximum volume only at ~85% of final dry weight. Borrás et al. suggest that for soya bean, which matures later in the year, reaching maximum seed volume later extends the capacity of soya bean to increase seed weight.

In iteroparous plants there is considerable variety in the growth process. For bristlecone pine physiological and anatomical properties, particularly of foliage, enable growth in a severe environment that seems likely to continue as long as foliage reiteration occurs and new branches have a water supply. In Douglas fir increasing tree height results in a decline in water potential and so an increase

in the possibility of xylem embolisms that is mitigated, in part, by anatomical changes in the new xylem produced. The continued production of new foliage through epicormic branches tends to produce foliage with a more favourable water status. In both species the morphological processes of reiteration determine a limit to growth.

In iteroparous plants decline and death is not prevented but delayed. The changes in physiology and/or structure that occur mitigate some of the effects of ageing but do not prevent decline, particularly where there is an increase in size. Magnani et al. (2008) suggest that in Scots pine (*Pinus sylvestris*) there are 'co-occurring changes in the balance between foliage, conducting sapwood and fine roots' so that water transport is maintained as plants grow taller and they refer to this process as 'functional homeostasis'. However, the anatomical and morphological changes that occur in Douglas fir as it ages do not maintain water status of the tree in the condition of the young tree—there is not a homeostatic condition—but they do reduce the decline from what may otherwise have occurred and above ground net primary productivity declines as plants age.

Similarly to semelparous plants the processes involved in ageing vary between species. Kennedy et al. (2010) compared branch growth and development of Douglas fir and grand fir (*Abies grandis*), co-occurring species in old-growth forests of North America that reach similar heights but Douglas fir can reach ~1,200 years while grand fir reaches ~500 years. These species have distinctive branch morphologies (Fig. 10.9).

Kennedy et al. show that while that of Douglas fir minimizes mean path length to terminal foliage, grand-fir minimizes the number of junction constrictions, and that the two morphologies result in different trade-offs between foliage display and hydraulic functioning. Douglas fir ecology involves longevity required to regenerate in the infrequent stand disturbances that enable its regeneration. Grand fir is a shade -tolerant tree and its ability to compete within an existing foliage canopy is aided by branch extension that occupies gaps that may occur.

Fig. 10.9 Contrasting crown structures of trees in an old-growth forest; left, western red-cedar, *Thuja plicata*; centre, grand fir, *Abies grandis*; right, Douglas fir, *Pseudotsuga menziesii*. Newly extended shoots are lighter green and illustrate branch and crown structures of *Abies* and *Pseudotsuga*. Foliage of *Abies* is held at the terminal section of branches; that of *Pseudotsuga* occurs in overlapping SCU. Photograph taken at the Wind River Canopy Crane Facility by Dr. M. C. Kennedy.

10.6 References

Acciaresi, H. A., Tambussi, E. A., Antonietta, M., Zuluaga, M. S., Andrade, F. H. and Guiamét, J. J. 2014. Carbon assimilation, leaf area dynamics, and grain yield in contemporary earlier- and later-senescing maize hybrids. *European Journal of Agronomy*, 59, 29–38.

Assmann, E. 1970. *The Principles of Forest Yield Study: Studies in the Organic Production, Structure, Increment, and Yield of Forest Stands*. Oxford, Pergamon.

Aumann, C. A. and Ford, E. D. 2006. Simulation of effects of wood microstructure on water transport. *Tree Physiology*, 26, 285–301.

Bégin, C. and Filion, L. 1999. Black spruce (*Picea marianna*) architecture. *Canadian Journal of Botany*, 77, 664–672.

Bond, B. J. 2000. Age-related changes in photosynthesis of woody plants. *Trends in Plant Science*, 5, 349–353.

Borrás, L. and Gambín, B. L. 2010. Trait dissection of maize kernel weight: towards integrating hierarchical scales using a plant growth approach. *Field Crops Research*, 118, 1–12.

Borrás, L., Slafer, G. A. and Otegui, M. E. 2004. Seed dry weight response to source–sink manipulations in wheat, maize and soybean: a quantitative reappraisal. *Field Crops Research*, 86, 131–146.

Borrell, A., Hammer, G. and van Oosterom, E. 2001. Staygreen: a consequence of the balance between supply and demand for nitrogen during grain filling? *Annals of Applied Biology*, 138, 91–95.

Brooks, J. R., Schulte, P. J., Bond, B. J., et al. 2003. Does foliage on the same branch compete for the same water? Experiments on Douglas-fir trees. *Trees: Structure and Function*, 17, 101–108.

Ciampitti, I. A. and Vyn, T. J. 2014. Understanding global and historical nutrient use efficiencies for closing maize yield gaps. *Agronomy Journal*, 106, 2107–2117.

Cliquet, J.-B., Deléens, E. and Mariotti, A. 1990. C and N mobilization from stalk and leaves during kernel filling by ^{13}C and ^{15}N tracing in *Zea mays* L. *Plant Physiology*, 94, 1547–1553.

Crawford, Jr., T. W, Victor, V. R. and Francis, E. B. 1982. Sources, fluxes, and sinks of nitrogen during early reproductive growth of maize (*Zea mays* L.). *Plant Physiology*, 70, 1654–1660.

de Kroon, H., Visser, E. J. W., Huber, H., Mommer, L. and Hutchings, M. J. 2009. A modular concept of plant foraging behaviour: the interplay between local responses and systemic control. *Plant Cell and Environment*, 32, 704–712.

de Witte, L. C. and Stöcklin, J. 2010. Longevity of clonal plants: why it matters and how to measure it. *Annals of Botany*, 106, 859–870.

Domec, J. C. and Gartner, B. L. 2002. Age- and position-related changes in hydraulic versus mechanical dysfunction of xylem: inferring the design criteria for Douglas-fir wood structure. *Tree Physiology*, 22, 91–104.

Domec, J. C. and Gartner, B. L. 2003. Relationship between growth rates and xylem hydraulic characteristics in young, mature and old-growth ponderosa pine trees. *Plant Cell and Environment*, 26, 471–483.

Domec, J. C., Lachenbruch, B., Meinzer, F. C., Woodruff, D. R., Warren, J. M. and McCulloh, K. A. 2008. Maximum height in a conifer is associated with conflicting requirements for xylem design. *Proceedings of the National Academy of Sciences of the United States of America*, 105, 12069–12074.

Drake, J. E., Raetz, L. M., Davis, S. C. and Delucia, E. H. 2010. Hydraulic limitation not declining nitrogen availability causes the age-related photosynthetic decline in loblolly pine (*Pinus taeda* L.). *Plant Cell and Environment*, 33, 1756–1766.

Flanary, B. E. and Kletetschka, G. 2005. Analysis of telomere length and telomerase activity in tree species of various life-spans, and with age in the bristlecone pine *Pinus longaeva*. *Biogerontology*, 6, 101–111.

Gaillard, J. M. and Lemaître, J. F. 2020. An integrative view of senescence in nature. *Functional Ecology*, 34, 4–16.

Gambín, B. L., Borrás, L. and Otegui, M. E. 2006. Source–sink relations and kernel weight differences in maize temperate hybrids. *Field Crops Research*, 95, 316–326.

Harley, C. B., Vaziri, H., Counter, C. M. and Allsopp, R. C. 1992. The telomere hypothesis of cellular aging *Experimental Gerontology*, 27, 375–382.

Hubbard, R. M., Bond, B. J. and Ryan, M. G. 1999. Evidence that hydraulic conductance limits photosynthesis in old *Pinus ponderosa* trees. *Tree Physiology*, 19, 165–172.

Ishii, H. and Ford, E. D. 2001. The role of epicormic shoot production in maintaining foliage in old *Pseudotsuga menziesii*. *Canadian Journal of Botany*, 79, 251–264.

Ishii, H. T., Ford, E. D. and Kennedy, M. C. 2007. Physiological and ecological implications of adaptive reiteration as a mechanism for crown maintenance and longevity. Tree Physiology, 27, 455–462.

Ishii, H. T., Jennings, G. M., Sillett, S. C. and Koch, G. W. 2008. Hydrostatic constraints on morphological exploitation of light in tall *Sequoia sempervirens* trees. *Oecologia*, 156, 751–763.

Kennedy, M. C., Ford, E. D. and Hinckley, T. M. 2010. Defining how aging *Pseudotsuga* and *Abies* compensate for multiple stresses through multi-criteria assessment of a functional–structural model. *Tree Physiology*, 30, 3–22.

Koch, G. W., Sillett, S. C., Jennings, G. M. and Davis, S. D. 2004. The limits to tree height. *Nature*, 428, 851–854.

Körner, C. 2003. Tansley Lecture: carbon limitation in trees. *Journal of Ecology*, 91, 4–17.

Kosgey, J. R., Moot, D. J., Fletcher, A. L. and McKenzie, B. A. 2013. Dry matter accumulation and post-silking N economy of 'stay-green' maize (*Zea mays* L.) hybrids. *European Journal of Agronomy*, 51, 43–52.

Lanner, R. M. and Connor, K. F. 2001. Does bristlecone pine senesce? *Experimental Gerontology*, 36, 675–685.

Larson, D. W., Doubt, J. and Matthessears, U. 1994. Radially sectored hydraulic pathways in the xylem of *Thuja occidentalis* as revealed by the use of dyes. *International Journal of Plant Sciences*, 155, 569–582.

Larson, D. W., Matthessears, U. and Kelly, P. E. 1993. Cambial dieback and partial shoot mortality in cliff-face *Thuja occidentalis*—evidence for sectored radial architecture. *International Journal of Plant Sciences*, 154, 496–505.

Lemoine, R., La Camera, S., Atanassova, R., et al. 2013. Source-to-sink transport of sugar and regulation by environmental factors. *Frontiers in Plant Science*, 4, Article 272.

Lim, P. O., Kim, H. J. and Nam, H. G. 2007. Leaf senescence. *Annual Review of Plant Biology*, 58, 115–136.

Ma, B. L. and Dwyer, L. M. 1998. Nitrogen uptake and use of two contrasting maize hybrids differing in leaf senescence. *Plant and Soil*, 199, 283–291.

Magnani, F., Bensada, A., Cinnirella, S., Ripullone, F. and Borghetti, M. 2008. Hydraulic limitations and water-use efficiency in *Pinus pinaster* along a chronosequence. *Canadian Journal of Forest Research*, 38, 73–81.

McCulloh, K. A., Domec, J. C., Johnson, D. M., Smith, D. D. and Meinzer, F. C. 2019. A dynamic yet vulnerable pipeline: integration and coordination of hydraulic traits across whole plants. *Plant Cell and Environment*, 42, 2789–2807.

Medawar, P. B. 1952. *An Unsolved Problem of Biology: An Inaugural Lecture Delivered at University College, London, 6 December 1951.* London, H. K. Lewis for University College London.

Meier, A. R., Saunders, M. R. and Michler, C. H. 2012. Epicormic buds in trees: a review of bud establishment, development and dormancy release. *Tree Physiology*, 32, 565–584.

Meinzer, F. C., Bond, B. J. and Karanian, J. A. 2008. Biophysical constraints on leaf expansion in a tall conifer. *Tree Physiology*, 28, 197–206.

Molisch, H. 1938. *Die Lebensdauer der Pflanze [The Longevity of Plants]. Translator E. Fulling.* New York, Science Press.

Moss, D. N. 1962. Photosynthesis and barrenness. *Crop Science*, 2, 366–367.

Muller, B., Pantin, F., Genard, M., et al. 2011. Water deficits uncouple growth from photosynthesis, increase C content, and modify the relationships between C and growth in sink organs. *Journal of Experimental Botany*, 62, 1715–1729.

Niinemets, U. 2002. Stomatal conductance alone does not explain the decline in foliar photosynthetic rates with increasing tree age and size in *Picea abies* and *Pinus sylvestris*. *Tree Physiology*, 22, 515–535.

Oenema, O. 2015. *Nitrogen Use Efficiency (NUE): An Indicator for the Utilisation of Nitrogen in Agricultural and Food Systems.* Colchester, International Fertiliser Society.

Ordóñez, R. A., Savin, R., Cossani, C. M. and Slafer, G. A. 2018. Maize grain weight sensitivity to source-sink manipulations under a wide range of field conditions. *Crop Science*, 58, 2542–2557.

Pan, W. L., James, J. C., Jackson, W. A. and Robert, H. M. 1986. Utilization of previously accumulated and concurrently absorbed nitrogen during reproductive growth in maize: influence of prolificacy and nitrogen source. *Plant Physiology*, 82, 247–253.

Petit, G., Pfautsch, S., Anfodillo, T. and Adams, M. A. 2010. The challenge of tree height in *Eucalyptus regnans*: when xylem tapering overcomes hydraulic resistance. *New Phytologist*, 187, 1146–1153.

Rajcan, I. and Tollenaar, M. 1999a. Source: sink ratio and leaf senescence in maize: I. Dry matter accumulation and partitioning during grain filling. *Field Crops Research*, 60, 245–253.

Rajcan, I. and Tollenaar, M. 1999b. Source: sink ratio and leaf senescence in maize: II. Nitrogen metabolism during grain filling. *Field Crops Research*, 60, 255–265.

Roach, D. A. and Smith, E. F. 2020. Life-history trade-offs and senescence in plants. *Functional Ecology*, 34, 17–25.

Rosbakh, S. and Poschlod, P. 2018. Killing me slowly: harsh environment extends plant maximum life span. *Basic and Applied Ecology*, 28, 17–26.

Ryan, M. G., Binkley, D. and Fownes, J. H. 1997. Age-related decline in forest productivity: pattern and process. *Advances in Ecological Research*, 27, 213–262.

Ryan, M. G., Binkley, D., Fownes, J. H., Giardina, C. P. and Senock, R. S. 2004. An experimental test of the causes of forest growth decline with stand age. *Ecological Monographs*, 74, 393–414.

Ryan, M. G. and Waring, R. H. 1992. Maintenance respiration and stand development in a sub-alpine lodgepole pine forest. *Ecology*, 73, 2100–2108.

Sala, A. and Hoch, G. 2009. Height-related growth declines in ponderosa pine are not due to carbon limitation. *Plant, Cell and Environment*, 32, 22–30.

Schulte, P. J. and Brooks, J. R. 2003. Branch junctions and the flow of water through xylem in Douglas-fir and ponderosa pine stems. *Journal of Experimental Botany*, 54, 1597–1605.

Sendall, K. M. and Reich, P. B. 2013. Variation in leaf and twig CO_2 flux as a function of plant size: a comparison of seedlings, saplings and trees. *Tree Physiology*, 33, 713–729.

Taiz, L., Zeiger, E., Møller, I. M. and Murphy, A. 2022. *Plant Physiology and Development*, Seventh Edition. Oxford, Oxford University Press.

Thomas, H. 2013. Senescence, ageing and death of the whole plant. *New Phytologist*, 197, 696–711.

Thomas, S. C. 2010. Photosynthetic capacity peaks at intermediate size in temperate deciduous trees. *Tree Physiology*, 30, 555–573.

Tyree, M. T. and Ewers, F. W. 1991. The hydraulic architecture of trees and other woody plants. *New Phytologist*, 119, 345–360.

Winner, W. E., Thomas, S. C., Berry, J. A., et al. 2004. Canopy carbon gain and water use: analysis of old-growth conifers in the Pacific Northwest. *Ecosystems*, 7, 482–497.

Woodruff, D. R., Bond, B. J. and Meinzer, F. C. 2004. Does turgor limit growth in tall trees? *Plant, Cell and Environment*, 27, 229–236.

Yang, L., Guo, S., Chen, Q. W., Chen, F. J., Yuan, L. X. and Mi, G. H. 2016. Use of the stable nitrogen isotope to reveal the source–sink regulation of nitrogen uptake and remobilization during grain filling phase in maize. *PLoS ONE*, 11, e0162201.

Yoder, B. J., Ryan, M. G., Waring, R. H., Schoettle, A. W. and Kaufmann, M. R. 1994. Evidence of reduced photosynthetic rates in old trees. *Forest Science*, 40, 513–527.

Zimmermann, M. H. 1983. *Xylem Structure and the Ascent of Sap.* Berlin, Springer.

CHAPTER 11

Conclusion

11.1 The meaning of 'dynamic' when applied to plant growth

Plant growth progresses through interactions between changes in the environment, plant development, structural progression, and physiological processes. These interactions form a system with four characteristics that together define the meaning of 'dynamic' when applied to plant growth. These characteristics determine how the plant growth process can be studied and how information about it can be represented.

11.1.1 There is continuous change in environmental conditions affecting growth

There are two types of environmental change: those external and those internal to the plant. External changes are most apparent as the result of seasonal, diurnal, and weather-driven variation in radiation, temperature, and moisture but can also be part of plant interactions with external conditions—for example, transpiration-induced reduction in soil moisture (Chapter 9). The plant microclimate is changed by plants in a number of ways as they grow.

Internal changes are inevitable as plants increase in size, perhaps most particularly in plant water status as height increases (Chapter 10). However, the magnitude of change is likely to depend upon the actual size reached. Change also occurs due to the increasing separation of tissues that have different but interdependent functions—foliage and roots being an obvious example.

11.1.2 Plants have control systems that respond to variation

Structural features are initiated through expression of genetics during development, and the ability to respond to variation can be a property of the structure and composition of the feature. An example is the violaxanthin ↔ antheraxanthin ↔ zeaxanthin photoprotection system (Chapter 6) which depends upon the structure of these molecules, that of the chlorophyll molecules they interact with, the spatial arrangement between them, and properties of the epoxidase/de-epoxidase enzymes involved in molecular changes. The components of that protection system are continuously present and it can be considered passive control.

Active control occurs when there is nuclear involvement in producing a response to environmental variation through new or changed structure. The circadian clock is an example that governs multiple components of the growth process (Chapter 5). A continuous biochemical oscillator is the foundation of this control, so it might be thought of as providing passive control, but the crucial feature making this an active process is the active sensing of daylight hours that results in changes in expression of some genes.

Plant control systems typically have distinct domains of action that define when, and to what degree, the control may affect growth rates. For example, processes controlled by the circadian clock (Chapter 5) are restricted to a particular segment of the diurnal cycle. Similarly, the annual cycle of perennial development, notably the establishment and then release from dormancy (Fig. 9.5), defines

The Dynamics of Plant Growth. E. David Ford, Oxford University Press. © E. David Ford (2023). DOI: 10.1093/oso/9780192867179.003.0011

domains of action for a number of growth processes. Plasticity and acclimation are induced by particular events or conditions which define their domains of activity.

Interactions between component processes may also have controls, although, perhaps not surprisingly, these seem to be less studied than the processes themselves. For example, the adaptive reiteration of branches in *Pseudotsuga menziesii* crowns (Chapter 10) involves production of epicormic buds in newly forming shoots, and then a delay of a number of years before the epicormic bud sprouts to form a new shoot. Although the effect of adaptive reiteration may be on the water status of growing tissue, this system can be understood as the result of two processes—epicormic bud initiation and suppression of growth due to a form of apical dominance, which is subsequently released—or some additional process may be involved. The shoot system that then forms likely has higher water potential than shoots at the further extension of the branch that produced the buds (Fig. 10.8).

There are also important spatial components to this process. Epicormic buds are on the adaxial surface of the branch and typically at the base of a lateral branch initiated immediately proximal to the parent branch apical bud. Visual observations suggest that the delayed epicormic shoot sprouting occurs when the apex of the lateral branch ceases to produce a new annual shoot and this new shoot system forms above the previous one and so is not shaded by it.

11.1.3 The history of a plant's development and growth affects its future growth

Components of growth are sequentially dependent with the development process controlling the initiation of new growth activity (Chapter 5). The sequential relationship is not necessarily representable as a single event but typically as a repeating progression as new meristems, and/or modules. This repetition produces both temporal and spatial separation of some processes and exploration and utilization of the environment which future growth depends upon as well as extending the separation of tissues with essential functions, e.g. the effect of increasing

plant height on water status (Chapter 10). The structural developments of ageing plants *Pinus longaeva*, *Zea mays*, and *Pseudotsuga menziesii* are examples of the dependence of growth on previous structure.

11.1.4 Genetic variability within a species can produce differences in both the amount and pattern of growth

Control systems of growth processes are under genetic control and consequently are subject to selection, whether natural or by humans as in the example of crop breeding (Chapter 2). In natural populations of a plant species growth processes vary between individuals. The differences between clones of *Picea sitchensis* (Chapter 7) illustrate that variation can be considerable both in quantity and plant structure.

In breeding programmes such as that for *Zea mays* (Chapter 2) breeders maintain multiple lines, each with particular characteristics, through inbreeding that maintains their genetic uniformity. Hybrids are produced from these lines for commercial use in production by selecting inbred parents with what are considered characters to confer advantages. The complete genetic structure of a species enables some production of viable individuals in response to environmental change or change in growing conditions, and in this sense genetic structure of the population is the complete repository of growth dynamics. The evolution of plants has enabled their continued existence in variable and changing conditions.

11.2 Study of plant growth as a dynamic process

Even from the small number of species discussed in this book it is clear that there is great variation between them in the amount and pattern of growth. Nevertheless, there is consistency between them in the processes involved and a *standpoint* can be defined and used for study of the growth of a particular species, or possibly some variant. This maintains that four characteristics of plants, as dynamic systems, must be considered to define how growth is controlled in a particular instance.

11.2.1 Standpoint

The standpoint has the following four components:

1. *Growth is the result of a sequence of processes that, in broad terms, starts with developmental processes involved in meristem activity and continues with structural development, autotrophic processes, and then synthesis of material for the plant body.*

   The sequence of processes is consistent between species but their duration may vary, as may the timing in the complete life cycle. In many tree species shoot extension and foliage production are restricted to limited periods (Chapter 9). In annual plants, such as *Zea mays*, vegetative growth may be separated in time from growth of reproductive organs.

2. *The environment, whether internal to the plant or external, may affect each of these processes differently.*

   An example illustrating contrasting effects is provided by the photosynthesis process of *Abies amabilis* which can be affected both positively and negatively by radiation (Chapters 3 and 6). Shoot extension in *Picea sitchensis* is likely affected by plant water potentials as they vary over the day (Chapter 9). It is important not to *assume* that different processes respond to variation in an environmental factor in the same way.

3. *It is necessary to understand interactions between processes in order to explain how growth is controlled.*

   Plant growth is the result of multiple interacting processes. Only in restricted experimental conditions or intensive and controlled agriculture is it likely that differences in growth can be attributed to one factor. Examples throughout this book illustrate that growth or essential processes that contribute to growth are the result of interactions between processes. Defining growth as a dynamic process acknowledges the importance of these interactions and of the role of change in the external and internal environments.

   Shoot extension (Chapter 9) provides an example of multiple interacting processes. It is reported to be affected by water status which varies diurnally as well as over an extended period of days. This contributes some understanding to its variation, with biological processes involved in both recovery from water loss and the response of tissues. However, extension is also affected by the developmental stage of the plant, increasing to a seasonal maximum and then decreasing, a pattern that does not appear a simple reflection of an environmental factor likely involved in a mechanistic process. An understanding of shoot extension requires that both sets of processes need to be taken into account: those associated with water status along with the effects of the development process—neither has priority in providing an explanation of variation in extension.

   Although it is limited in scope (what should be considered) and detail (how the component controls function) the system represented in Fig. 9.13 provides some understanding of the variation that can be found in shoot extension, due to weather-related and developmental processes. A more complete understanding could be obtained through study of the interactions between the cell water uptake → cell expansion process and developmental processes both seasonal and diurnal and possibly additional processes not represented in Fig. 9.13. For example, there is a decline in extension between 00:00 and 04:00 h, but this reduction is greater following a day of greater day-time shoot contraction (Chapter 9) even though shoots can be expected to have recovered a high water potential during the 00:00 and 04:00 h period. Is this decline generally due to a developmental process or possibly related to gradual depletion of carbohydrate resources for growth? And why should it change following a day of less extension? Greater understanding would require investigation of component processes and particularly their interactions.

4. *Primary control of development of the plant body is dispersed; it occurs within each apical module of the plant where genetic programmes interact with availability of resources and environmental conditions and this may result in expression of plasticity. The way that modules are integrated to form what appears as the complete plant is the result of this dispersed system of control.*

   Plant structure has an effect on the interaction between the plant and its external environment, as in the interception of light and its potential effect on photosynthesis for the *Zea mays* canopy

(Chapter 8). The structural pattern that a plant follows during growth may also affect how much growth can be made, as in the case of clonal differences in *Picea sitchensis* (Chapter 7) where differences in production of buds of different types, their extension, and foliage amount are all related to the amount of growth made. The architecture of a plant follows a genetic blueprint, but its complete realization depends upon available resources and environmental conditions at the module that generates the architectural development (Chapter 7).

11.2.2 Integration of the plant body

In the study of growth, say of a particular species, a central question is:

Which process or processes have a controlling effect on growth in response to change?

An answer, say for a particular plant in a specified environment, requires determining the importance of processes of different types and how they interact in producing growth.

A challenge is that while growth is the result of interactions between control processes within plants, there is no central controller. This contrasts with many animals that have central nervous and circulation systems which together coordinate activities of distributed organs with different functions in bodies that remain comparatively fixed in size once an initial growth period terminates. An essential feature of most higher plants is that from a fixed position they maintain growth which enables continued resource acquisition from both below and above ground. This activity depends upon the development and spatial extension of apical meristems, and the construction of organs from them that collect or intercept resources for the autotrophic process.

Consequently, for plant growth, the term *integrated* requires a unique definition. With regard to systems in general, *integrate* means 'an organized array of individual elements and parts forming and working as a unit'. Plants are integrated in the sense of an expanding form that acquires resources from the environment and is organized by having connections between successive modules but is not

an integrated system in the sense of there being a central organizing process. Physical connection between component modules across the network of a plant body does not imply a unified physiological dependence across the complete network of the plant: modules may senesce and die and, in some cases, may even develop into separate new plants. The characteristics of integration through connections between modules need to be researched for plant types and species.

11.2.3 Species-specific approach

The components of growth require detailed descriptions for individual species and investigations of their relative importance. Because species vary in their development and morphology, as well as possibly in interactions between processes, these descriptions will be different.

The first requirement is to establish the continuing interactions between development, morphological structure, and foliage amount and its distribution that result in the expansion of the plant. The models developed by Hallé et al. (1978) provide an understanding of core developmental and morphological processes that can be involved, but their use in the analysis of growth requires extension to include foliage and to specify the times when meristems are active.

As an example Fig. 2.13 provides an outline for *Zea mays* that incorporates four control processes involved in module growth and the critical developmental transition from vegetative to reproductive growth. Interactions between environmental conditions and these processes are described in Chapter 2. More detail is given for canopy development and functioning in Chapter 8 and for the reproductive phase in Chapter 10.

Figure 2.8 provides an outline of the critical foliage and shoot regeneration process of *Pinus longaeva* and that has a continuing cyclical development in contrast to the linear development of the annual *Zea mays*. Foliage structure and foliage longevity and their roles are described in Chapter 2.

In both examples an essential feature is to represent how change occurs in the development of plant form and the effects that it has on growth, whether induced by a developmental process in *Zea mays* or perturbation affecting plant structure in *Pinus longaeva*. Interactions between processes are clearly different for the two species.

In some situations distinctions in the growth process may be for particular components. For example, in the case of the morphological component of *Picea* species (Fig. 7.6), which all have the same basic architecture, there are differences in foliage size and distribution on the shoot and in shoot rigidity.

For larger plants the individual can be the result of competition between apices for resources, light, and water/nutrition, and hormonal control. Modules, or in trees collections of modules organized as branches, can senesce, which is an active process typically related to a change in their environment that reduces capacity to obtain and/or utilize resources for growth. Foliage, and particularly branches of some trees, can function as at least partially independent (e.g. Sprugel 2002), with much of the control of growth being local to the module(s). Physical support of leaves by a stem or branches by the main trunk obviously affects their environment and so their function.

11.3 Causality

Investigation of causal relationships can be considered as an antidote to making premature conclusions based on correlation (Box 1.1), where a cause may involve a transfer of mass, energy, or signals that influence a process, or a change in conditions. Seeking cause is important where contrastive questions of the form 'Why this and not that?' are asked.

Seeking causal relationships in science has been considered an ideal, often implicitly. However, in the study of plant growth there are complications. Genetics determines the structure and method of action of contributing processes to growth. There are multiple processes and so multiple genes involved and this is found in quantitative trait loci investigations (Chapter 1). Furthermore, a growth effect may be achieved in different ways. Investigation of *Picea sitchensis* clones (Chapter 7) showed variation between module size, density of foliage along modules, and crown structure that can affect illumination and may each have an effect on growth,

and different combinations of these characters may result in the same or similar growth.

Growth is the result of interactions, so that it is unlikely for there to be a single cause of why one plant may grow more than another, except perhaps in extreme environmental conditions. What needs to be understood are how interactions between processes take place. This does require an understanding of the processes themselves.

11.4 Quantitative expression of biological phenomena

For some scientists, questions involving plant growth require quantitative answers and modelling of various types is widely practised. However, dissatisfaction has been expressed about the approaches and achievements of modelling. Tardieu (2010) in editorial comments to a collection of modelling papers stated:

Plant modelling has not yet become a standard method in Plant Biology for integrating complex mechanisms, nor in Genetics for designing new plants. This contrasts with the situation in industry in which, for example, modelling has in good part replaced costly experiments in wind tunnels for the design of cars or planes.

He notes that a model for plant performance of an 'average genotype' may be effectively calibrated for a set of well-known experimental scenarios. But for detailed questions representations of plant processes are subject to debate 'because they are not based on uncontroversial physical laws' and he suggests the nature of the subject means that the term 'mechanistic model' is unfortunate in plant modelling. Recall Hunt's (1982) conclusion that growth is a 'mathematically form-free process'. A theory is required before a quantitative model can be constructed for a specific case and the work reported in this book suggests that for many cases considerable research is still required in developing theories that are adequate as a basis for modelling. The causal relationship on which plant growth rests is the genetic structure of the plant and the very essence of genetic control is that it is variable and this is always likely to present a challenge to modelling.

Models may be formed by making simplifying assumptions about some component process.

Modellers make choices about the mathematical function to be used to express relationships. Choices are frequently guided by an ideal of simplicity, as in the growth curves described in Chapter 1, or because a well-known branch of mathematics can be used, e.g. differential calculus. Such models can become enshrined in a subject, have widespread use, and can restrict progress. Plant growth processes are determined by a plant's genetics. Variation in component processes and their interactions can be expected and is what needs to be understood.

The programme of functional–structural plant modelling (FSPM) developed to integrate plant structure into an understanding of the control of plant growth. Research includes construction of species-specific modelling platforms, for example, GREENLAB (Yan et al. 2004) and LIGNUM (Perttunen et al. 1998). These model approaches use a basic framework for a species that can be used to express variation in other components. LIGNUM can be considered as a discrete event simulator for time and morphological position on the plant, while GREENLAB is based upon an automaton-based simulation of organogenesis.

The FSPM programme has involved investigation of many components and examples of plant growth and has been reported at a succession of meetings. The first such meeting was held in Finland (Anon. 1997). Louarn and Song (2020) review two decades of work.

Vos et al. (2010) define FSPM as follows:

Plants react to their environment and to management interventions by adjusting physiological functions and structure. Functional–structural plant models (FSPM) combine the representation of three-dimensional (3D) plant structure with selected physiological functions. An FSPM consists of an architectural part (plant structure) and a process part (plant functioning). The first deals with (i) the types of organs that are initiated and the way these are connected (topology), (ii) co-ordination in organ expansion dynamics, and (iii) geometrical variables (e.g. leaf angles, leaf curvature). The process part may include any physiological or physical process that affects plant growth and development (e.g. photosynthesis, carbon allocation).

FSPM, as defined by Vos et al. can be calibrated for use in agriculture (Vos et al. 2007). Much FSPM research has focussed on developing quantitative

description of plant morphology and its application, such as L-systems (e.g. Prusinkiewicz et al. 2018), but the separation of the architectural part of a model from the process part may limit use in study of the growth process, particularly investigation of interactions between processes with different functions.

In their review of two decades of FSPM research Louarn and Song (2020) note achievements in development of representations of both plant architectures in space and time and scheduling of organ production and its physiology. But they suggest that for FSPM:

After a mainly exploratory phase, the approach has yet to reach its full potential in terms of integration and heuristic knowledge production, but its coverage of the full continuum—from elementary traits to complex plant phenotype and then population functioning renders it a unique tool for modern biology ... This places it in a central position at the crossroads of fundamental questions in plant biology and predictive ecology.

In particular they suggest:

the full promise of FSPMs is still far from being delivered, even after two decades. One striking observation is that until now the vast majority of studies have focused on particular biological questions, using models applied specifically to a single species or integrating a limited array of physiological processes. The ability to effectively derive knowledge from the integration of complex structures and functions is thus likely still in its infancy.

The results reported in the chapters of this book are largely from biological investigations and the suggested standpoint to be used in developing an understanding of growth is expressed verbally. This is not surprising. Although quantitative approaches may be used in particular studies, the body of research on which it is based is developmental, genetic, and physiological, and understanding of these subjects does not *depend* upon quantitative models.

11.5 References

Anon. 1997. Special issue on functional–structural tree models. *Silva Fennica*, 31, 239–380.

Hallé, F., Oldeman, R. A. A. and Tomlinson, P. B. 1978. *Tropical Trees and Forests: An Architectural Analysis*. New York, Springer.

Hunt, R. 1982. *Plant Growth Curves. The Functional Approach to Plant Growth Analysis*. London, Edward Arnold.

Louarn, G. and Song, Y. H. 2020. Two decades of functional–structural plant modelling: now addressing fundamental questions in systems biology and predictive ecology. *Annals of Botany*, 126, 501–509.

Perttunen, J., Sievänen, R. and Nikinmaa, E. 1998. LIGNUM: a model combining the structure and the functioning of trees. *Ecological Modelling*, 108, 189–198.

Prusinkiewicz, P., Cieslak, M., Ferraro, P. and Hanan, J. 2018. Modeling plant development with L-Systems. in: Morris, R. J. (ed.) *Mathematical Modelling in Plant Biology*, pp. 139–169. Cham: Springer International Publishing.

Sprugel, D. G. 2002. When branch autonomy fails: Milton's law of resource availability and allocation. *Tree Physiology*, 22, 1119–1124.

Tardieu, F. 2010. Why work and discuss the basic principles of plant modelling 50 years after the first plant models? *Journal of Experimental Botany*, 61, 2039–2041.

Vos, J., Evers, J. B., Buck-Sorlin, G. H., Andrieu, B., Chelle, M. and De Visser, P. H. B. 2010. Functional–structural plant modelling: a new versatile tool in crop science. *Journal of Experimental Botany*, 61, 2101–2115.

Vos, J., Marcelis, L. F. M., Visser, P. H. B, Struik, P. C. and Evers, J. B. (eds). 2007. *Functional–Structural Plant Modelling in Crop Production. Wageningen UR Frontis Series, Vol. 22.* Dordrecht, Springer.

Yan, H.-P., Kang, M. Z., De Reffye, P. and Dingkuhn, M. 2004. A dynamic, architectural plant model simulating resource-dependent growth. *Annals of Botany*, 93, 591–602.

Glossary

A glossary is an explanation of terms and so, for some, they are described as part of the process to which they contribute. The letter C followed by a number indicates a chapter in which the term is discussed.

AAA+ enzyme: see enzyme.

abscissic acid: see phytohormones.

acclimation C1: Development of an anatomical structure or property of a physiological process in response to an environmental condition. Typically found where there is variation in environmental conditions such as the environmental gradients as in a **canopy** with development of **sun** and **shade** foliage.

acid growth hypothesis C5: The hypothesis that acidification of the developing primary cell wall due to proton movement across the plasma membrane results in stress relaxation of the wall enabling its extension.

active control: see control processes.

advective cooling and transport: see microclimate.

ageing C10: To remain alive plants must maintain active meristems. This results in increase in size or a developmental change that affect vegetative growth and consequently longevity of the plant body. Typically growth in size of **iteroparous** plants can result in change in water supply to growing tissue as plants grow taller, and/or change in availability of resources for growth. While there are typically processes that mitigate these effects to some extent growth can be slowed. **Semelparous** plants have a developmentally produced change from vegetative to reproductive growth and there are processes that provide reproductive meristems with resources from the vegetative body that has ceased growing. **'Death by starvation'** is a theory that iteroparous plants, particularly trees, die because of reduced production of materials necessary for new growth. This is considered due to an increase of materials required to sustain the existing plant body as it increases in size and consequently a decline in production of foliage. This has largely been superseded by more integrated theory of environmental effects as size increases.

apex: see form.

apoplastic C9: The apoplast is the volume of a tissue that is external to plasma membranes and consists of cell walls, xylem, and intercellular spaces. Apoplastic transport is movement of molecules through this space.

A/Q curve C3: see photosynthesis.

aquaporins C4: Protein channels embedded in cell membranes that are water selective and provide faster diffusion than that directly through the lipid bilayer of of the membrane, Aquaporins can be reversibly **gated**, ie between open and closed.

architecture C7: Plant architecture is defined as 'The visible morphological expression of the genetic blueprint of a tree at anytime' (Chapter 7). The purpose of defining architecture is to define how morphology and development affect growth. 23 **architectural models** have been described with the essential feature being that the same model applies to species not taxonomically close so that models represent 'a biological necessity' for functional and competitive disposition of organs in plant' under s particular set of conditions. Growth of the above-ground part of a plant and its resulting architecture involves repeated increments of **modules** of growth. Typically a module comprises a unit of support structure, the foliage it supports, and meristems that can develop to continue growth. Modules may vary between species in anatomical and morphological structure even in those with the same architectural model (C2). A **phytomer** (or **metamer**) has a more restricted definition than module being a node and internode on a stem, the foliage at that node, and its subtended bud if present.

auxanometer C9: An instrument for measuring extension growth of plants.

biochemical oscillator: see development.

biomass C7: The weight of biological material usually per unit of ground occupied, e.g. g.m^{-2}, dried (80°C). Measures of biomass are typically fractionated, e.g. living dead, plant, animal. The term also has a more general use in non-fossil fuel energy studies.

boundary layer conductance: see microclimate.

C$_2$-cycle: see photosynthesis.

callose C9: A complex carbohydrate deposited between plasma membrane and cell wall. It can occupy a complete cell volume and prevent transport of materials and/or seal the cell following damage.

canopy C8: A canopy is a horizontally continuous cover of foliage. For a planted stand, for example a crop, then as individual plants grow foliage from neighbours starts to overlap and the canopy forms. The amount of foliage is frequently calculated as **leaf area index** (L), the area of foliage above a unit of ground. Typically agricultural crops have L between 3 and 4 at maturity; mature evergreen forests can have L=10. **Above-ground competition** for light between neighbouring plants occurs when shading due to foliage in one plant's crown influences growth of a neighbour. Shoots and foliage can develop towards brighter light in a canopy resulting in **resource acquisition plasticity**; growth in the upper part of the crown develops towards more illuminated space that results in a change in structure, i.e. **plasticity**. This can result in **crown asymmetry** where the centroid of the foliage crown is displaced horizontally from the position defined by the main stem of the plant. Continuous resource acquisition plasticity driving competition in the upper canopy results in spatial equalisation of large and/or surviving plants so successful individuals are spatially evenly distributed. Light absorption by foliage varies according to wavelength with red light being preferentially absorbed and reduced in reflection and transmission. This decreases the ratio of red to far red light, R_{660nm}/FR_{730nm}, where R and FR indicate fluence rate in a 10 nm band centered on the given figure of the sub-text. A number of growth and development proceses respond to this change resulting in **morphogenetic plasticity**. Penetration of light into a canopy has frequently been measured at successive horizontal layers from the top of the canopy and models to fit constructed this typically assumes exponential decline, as found in a homogenous medium (Beer's Law) but lack of homogeneity in canopies has presented difficulties in both measurement and interpretation. The method of **nested radiosity** calculates interception and absorption based upon the 3D geometry and optical properties of foliagew and characteristics of radiation.

canopy conductance: see microclimate.

carbon centric approach: see scientific methods.

causal network: see scientific methods.

causal processes: see scientific methods.

cell C9: Plant cells have an enclosing cell wall composed mainly of **cellulose** that gives rigidity and structural support to the cell and the tissue it is part of. The wall is composed of a **primary cell wall** that that develops during the growth of the cell cytoplasm and a **secondary cell wall** that thickens the wall once the cell has reached its final size. Water uptake into the cell **vacuole** is required to maintain the turgor of a cell as the cytoplasm expands and to provide the force for cell expansion. Water and solutes are taken into the vacuole which is surrounded by a membrane, the **tonoplast**, within which are embedded channels by which this uptake is controlled.

cellulose synthesis: Cell wall components are synthesised in Golgi bodies within the cytoplasm and then, through secretory vesicles, are transferred through the cytoplasm enclosing plasma membrane toward the developing wall. Celloluse microfibrils are synthesised at the cell surface by membrane bound cellulose synthesase proteins. Synthesis of the primary cell wall also involves pectins that are gel-forming components within which cellulose is embedded and provide structure but also enable expansion of the cell (see also **acid wall hypothesis**).

chaperone enzyme: see enzyme.

chlorophyll fluorescence: see photosynthesis.

chloroplast C6: Chloroplasts are organelles contained within cells that are the site of **photosynthesis**. They have a double outer membrane separating them from the cytosol. Their content comprises a fluid system, the **stroma**, which contains **RuBisCo** along with other enzymes, and **thylakoid membranes**. These are double membranes that contain the **thylakoid lumen**, These membranes form two types of structure: **grana** that are stacks of circular discs with close contact between outer membranes of members of the stack, and **stroma** lamella which are unstacked and ramify in the stroma and connect between **grana** members. Thylakoid memembranes are connected with continuous lumen through the thylakoids of a chloroplast. Light harvesting complexes and components of electron transfer systems are embedded in the thylakoid membranes.

circadian clock: see development.

clade C7: A clade comprises an organism and all of its descendants. Clades are typically defined by DNA analysis of potential members to examine for similarities and differences and to estimate evolutionary developments.

clone C7: A clone is a group of genetically identical individuals. A plant clone can be produced through repeated asexual reproduction. Where this occurs then the individuals form a **genet**: an individual in such a group is a **ramet**. Plants with extending rhizomes, which are modified underground plant stems may produce new shoots that grow upward and above-ground and while the growth of this new shoot is first dependent upon connection through the rhizome to older shoots, typically such shoots become independent and so are ramets.

closed-loop control: see control processes.

competition: see canopy.

conservation of genes C1 and C5: conserved sequences of genes are identical or similar sequences of nucleic acids or proteins in different species, or within a genome. This conservation is considered to indicate maintenance by natural selection and some evolutionary relationship between the organisms.

control processes, C1 and C 4: A control process consists of a command, typically a change in a condition of the environment or within the plant in this work, an **operand** that is affected by this change resulting in an output. The way that the operand functions determines whether control is **active** or **passive**. In active control the operand has some change in its state, for example a gene that produces a protein, or induction of an enzyme to a functioning state. In passive control the operand exists in its functional state prior to the command and remains in that state through to control operation. Control may be open- or closed-loop. In closed loop control the output is monitored in some way and the control system adjusts to produce a desired output. This is typically achieved through a **setpoint**, a desired value that can be compared with the output. In open-loop control the output is not monitored: accurate control can nevertheless be obtained if the operand is well tuned to requirements. For example in a **regulator** two open-loop systems may operate with the output of one being the command to the other and vice versa. The effectiveness of such a system depends upon how the two operands are balanced. Plant systems differ from those in engineering in the active changes that can occur in the operands, for example, enzyme control systems can be **induced** into action or may have a **chaperone system**. **Membrane control systems** have embedded operands, such as channels that enable passage of molecules and their numbers and types can be under active genetic control.

convective cooling and transport: see microclimate.

convergent evolution C7: The separate development of the same or similar characters in taxonomically unrelated or distinctly related species existing in similar habitats. When this occurs it is frequently taken as indication of the functional value of the character.

correlation analysis: see scientific methods.

crown asymmetry: see canopy.

dark cycle: see photosynthesis.

death by starvatio n, see ageing.

decurrent C8: A crown with decurrent branching has weak apical dominance and tends to be rounded, typical of many angiosperm trees and counter to **excurrent** branching.

dichotomous key C7: Dichotomous keys are used in the identification of plant species. Successive paired questions are asked about the structure, appearance, or sometimes habitat of a plant with the answers leading to further question pairs. This proceeds in a series until the species is identified.

diffusion porometer C9: An instrument for estimating the stomatal resistance of foliage to water vapour transfer. Foliage is placed in a chamber and the rate of increase in humidity is measured and the flux of water transpired per unit of leaf area calculated.

distal: see form.

development C2: Growth for many plants is not a smoothly continuous process but is modulated by developmental changes, such as the production of new types of organs or changes in state as in the onset or release from dormancy. Such changes are under genetic control frequently stimulated by a **phytohormone**. A core component of developmental control is the **circadian clock**, a nuclear process whereby a sequence of genes that form the clock stimulate or inhibit each others actions in a sequence. The sequence operates as a **biochemical oscillator**. Gene expression involves nucleic acid transcription, transport of mRNA from nucleus to cytoplasm, and synthesis of proteins that may inhibit or initiate gene expression. Proteins are broken down by enzymatic proteolysis. The time this process takes determines the rate at which one gene can affect the operation of another and this typically causes oscillation in their effects. The action of some genes is affected by light and this causes the circadian clock to follow changes in daylength. Genes of the clock, or their products can initiate or terminate transcription of non-clock genes so that their operation is controlled, known as **gating**. This, in combination with light modulation of the clock can produce **seasonality** in action of genes.

domain (of influence) C9: The domain of influence of a factor or process are the conditions under which it causes or responds to change. For example, multiple factors may affect meristem growth but their domains of influence are restricted to when the meristem is not dormant.

down regulation: see gene expression.

dynamic process C1: A dynamic process can change its output or operation in response to change in its environment. This can include maintaining a constant output while external conditions vary, as in homeostasis, or changing its level of output due to induction or inhibition of some component. Typically, responses to a change in biological systems are not instantaneous but the response is delayed or **lagged** as existing components **acclimate** to conditions or new components are produced.

electron transport: see photosynthesis.

empirical studies: see scientific methods.

enzyme C6: A protein produced under genetic control that catalyses chemical reactions in the plant. There are different types of enzyme, for example, **RuBisCo** functions with RuBisCo activase (rca): the active sites of RuBisCo can become inactive and rca reactivates is them and is referrred to as the RuBisCo **chaperoneB.** In this example RuBisCo is the apoenzyme and rca is the holoenzyme. rca is an **AAA+** enzyme; an enzyme that acts as a mechanical motor driven by ATP.

epicormic: see form.

Evapotranspiration: see microclimate.

excurrent C8: A crown with excurrent branching is cone shaped and has strong apical dominance by the main leading apex.

extension C7: Extension growth requires expansion of individual cells driven by an increase in turgor and the concomitant relaxation of cell wall structure that enables incorporation of material into the wall. See also **acid growth hypothesis.**

feedback: see control processes.

feedforward control: see control processes.

ferredoxin: see photosynthesis.

form C2: Plant **form** encompasses the morphology of components structures and how these are organised in the plant body and so contribute to the complete structure of the plant. Components of form and the processes by which they are produced are sometimes defined in terms of how they contribute to the growth of the plant. **Extension** of the above-ground plant body takes place through accretion of **module(s)** at each apex with an active **meristem.** A module comprises a unit of support structure, e.g. stem or twig, and its attached foliage. Modules can have different structures: a **monopodium** is produced by **indeterminate growth,** i.e. with no natural breaks in growth; a **sympodium** comprises repeated increments that may pause in growth such as in production of an inflorescence, which is an example of **determinate growth,** and then continue growth through production of a further module. **Sylleptic growth** occurs when meristems do not havedifferent dormancy or resting phases: lateral buds that develop at the same time as the apical bud show sylleptic growth. **Proleptic growth** occurs when a bud has delayed growth relative to its subtending apex; this occurs with **epicormic buds** that may remain

dormant for a considerable time. Proleptic growth is frequently the foundation of **reiteration** where a new branch repeats the same structure as its parent but delayed in time. The development of the form of a plant is sometimes described as in two phases: **organogenesis,** production of new meristems, e,g. as in buds; **extension,** the growth of those meristems. Structures on a module that are closest to its point of origin are referred to as **proximal;** those closest to its apex are **distal.** The artangement of organs around the module is the result of the plant's **phyllotaxis** which can be of different types, e.g. spiral, alternate. Organs, particularly foliage, may be flattened in horizontally inclined modules referred to as **plagiotropic** whereas those of vertical modules are not flattened and are **orthotropic.**

gating: see development.

gene expression C5: The basis of **gene expression** involves the nuclear process of RNA production from gene DNA, modification and transport of that RNA to a ribosome in the cytoplasm where it functions in protein production. The DNA → RNA process requires the enzyme RNA polymerase and its initiation to produce RNA is controlled by a promoter region of the DNA and also requires the presence of transcription factors and gene regulatory proteins: transcription may be enhanced or repressed by such proteins. All steps of gene expression may be regulated. **Downregulation** refers to decrease in RNA or protein synthesis in response to a stimulus of cellular factor and **upregulation** is the opposite.

genet: see clone.

gibberellic acid (GA): see phytohormone.

grana: see chloroplast.

growth curve: see quantitative model.

Haloenzyme: see enzyme

homeostasis C4: the maintenance of a constant state of an organism or one of its components despite changes in the environment it experiences. This is frequently applied to internal conditions of an organism. For plants, where growth involves continuous change in structure, and so of tissues, this concept can be difficult to apply other than in a general sense to imply the capacity to respond to changes to maintain the plant and its function.

hybrid plants C8: Hybridisation is a procedure used in plant breeding where two parents are selected for breeding, each with qualities considered desirable. Typically these parents are from inbred lines, i.e. produced through repeated self-fertilisation, and so are considered pure lines but which tends to decrease vigour. Breeders maintain in-bred lines and have understanding of their previous performance in producing hybrids for use in crop production. Crossing between unrelated in-bred lines typically restores productivity.

hysteresis C4: A system exhibits **hysteresis** when the its response to a change in environment lags behind the rate of change in the environment. An example is the contraction and re-expansion of a growing plant tissue in response to a decrease followed by an increase in tisue water potential. For the same value of water potential the volume of the tissue is less during re-expansion than contraction due to the time taken for the biological processes to produce the re-expansaion.

induction: see photosynthesis.

integration: see scientific methods.

iteroparous C10: An iteroparous plant produces seed more than once in ts lifetime and typically may produce seed many times. See also semelparous.

Interwhorl: see form, C7.

Lag: see dynamics.

leaf angle: The angle of a leaf from the horizontal plane affects the intensity at which it intercepts radiation from a particular direction and this can be important for rate of photosynthesis and heat balance and so temperature. The azimuth direction of the leaf, for example whether it faces north or south and how this relates to the azimuth direction of the sun, influences the effect of leaf angle. For some plants **leaves are curved** so that conditions at its surface may vary along the length of the leaf.

leaf area index: see canopy.

light capture: see photosynthesis.

light harvesting complexes: see photosynthesis.

logistic growth curve: see quantitative model.

open-loop control: see control processes.

mechanistic theory: see scientific methods.

meristem C5: A **meristem** is a plant tissue comprised of cells capable of actively dividing where some cells produced maintain the capacity for further division whereas others differentiuate into cells that form the tissue of new organs and are specialized enabling particular function. There are three main meristems involved in vegetative growth: the **shoot apical meristem**, the **root apical meristem**, and the **cambium** that produces stem thickening in woody plants. The dividing cells are **pluripotent**, ie. capable of differentiating into cells that make up the tissues of new organs. Typically cells undergoing different development processes occur in separated regions of the meristem and the processes involved depend upon phytohormones or other control agents. In the shoot apical meristem size of the zone of actively dividing cells is regulated by the CLAVATA-WUSCHER gene system: development of lateral organs in a phyllotaxis pattern is controlled by the **phytohormone auxin**.

mesophyll conductance: see photosynthesis.

Membrane nuclear controls: see control systems.

microclimate C3, C8 and C9: Physical conditions that can affect plant growth and physiological processes, notably radiation, temperature, humidity and air circulation, are determined not only by weather but by the structure and functioning of plants themselves particularly when they grow in **canopies**. Interaction between weather conditions and plant and/or canopy structure results in a **microclimate**. An important characteristic of microclimates is that conditions vary with position both for plants growing individually and in canopies. This variation is greater for plants within canopies within which there can be substantial vertical gradients in conditions. The mass of a plant intercepts radiation: **direct radiation** from the sun varies in flux density, angle, and azimuth direction; **diffuse radiation** has greater flux density under cloudy than clear sky conditions, is received from all directions of the sky hemisphere. In addition to there being gradients of radiation, both vertical and horizontal depending on the structure of the plant or canopy there are gradients of temperature, humidity and air speed. Radiation heats the material of the plant/canoopy and is a driver of **evapotranspiration**, the loss of water from with the plant or from its surface if wet. Transfer from plant or canopy elements to the surrounding air can result in reduced air density and **convective cooling** and upward transport of heat and water vapour. **Advective cooling** occurs when wind moves air across and through a canopy. These transport processes are affected by the structure of the vegetation which influences **boundary layer conductance**, the rate of movement from the canopy space to the atmosphere above it.

module: see architecture.

monopodium: see form.

morphogenetic plasticity: see canopy.

multiple regression tree: see scientific methods.

nested radiosity: see canopy.

nuclear control: see development.

open-loop control: see control processes.

operand: see control processes.

organogenesis: see form.

orthotropic: see form.

passive control: see control processes.

peroxisomes C6: Peroxisomes are organelles within the cytoplasm involved in the C_2-cycle associated with photorespiration: they receive glycolate and glutamate from chloroplasts, produce glycine that passes to mitochondria, that in turn is processed in production of serine that returns to peroxisomes where it, in turn, is processed in production of glycolate that returns to chloroplasts. Peroxisomes also contain enzymes that convert hydrogen peroxide to water and oxygen.

phloem C10: Phloem is the tissue that conducts the products of photosynthesis or breakdown of storage such as starch, particularly sugars, throughout the plant: the process of translocation. It is comprised of specialised cells notably **sieve cells** (gymnosperms) or **sieve tube elements** (angiosperms) that have no nucleus and through which sap flows, and **albuminous cells** (gymnosperms) and **companion cells** (angiosperms) that are involved in **loading** and **unloading** of molecules that are translocated. Flow in the phloem requires build-up of pressure at the loading site due to water uptake. Transport of sugars to an from phloem may be **apoplastic** or **symplastic** depending upon species, but their loading into, and unloading from, the phloem are active processes.

photosynthesis C6: The synthesis of organic compounds from inorganic CO_2 utilizing energy from light involves two processes: (1) The **light reactions**, involving chlorophyll mediated disassociation of water molecules and production of protons into the thylakoid lumen used in synthesis of ATP, and activated electrons transferred across the thylakoid menbrane with potential to produce NADPH; (2) the **dark cycle** or **Calvin-Benson-Bassam cycle (CBB)**, utilisation of light reaction products in the incorporation of C from CO_2 into organic molecules involving the enzyme **ribulose-1,5-bisphosphate carboxylase/oxygenase (RuBisCo)**.

There are three important nodes of control in this process: (a) **light energy capture** and activation of chlorophyll associated with **light harvesting complex (LHCII)** and disassociation of water molecules by **Photosystem II (PSII)** and production of a **linear electron flow (LEF)** through a series of electron acceptors; (b) the further light energy capture and excitation of electrons by chlorophyll of **light harvesting complex I (LHCI)** resulting in reduction and production of NADPH by **Photosystem I (PSI)**. Under some conditions, rather than being involved in this reduction **cyclical electron flow (CEF)** occurs back to the chlorophyll coupled with proton pumping utilised in production of ATP; (c) RuBisCo utilises the 5-carbon sugar ribulose 1,5-Bisphosphate (RuBP) with carbon from CO_2 to produce two molecules of 3-phosphoglycerate that are reduced to glyceraldehyde-3-phosphate. From this sugars are produced for other metabolic reactions and RuBP is regenerated.

The photochemical processes of PSII and PSI result in negative excited-state redox potential. The word **redox** combines reduction and oxidation and these two processes occur simultaneously, not separately. In the photosynthesis system this occurs by transfer of electrons and the terminal acceptor of this potential is NADP+ via ferredoxin resulting in its conversion to NADPH. However, if production of this potential exceeds the capacity of ferredoxin to receive it then potentially damaging reactions may occur and, although the focus of photosynthesis studies and teaching has most frequently been carbon fixation, there are multiple associated processes concerned with **protection** against damage and the **repair** of such damage that may occur. While both protection and repair processes conserve the photosynthesis system they can reduce CO_2 fixation. Five processes need to be considered in protection and repair.

(i) **Photoprotection.** Interception and dissipation of visible radiation. At high flux density of light the spatial organisation of chlorophyll molecules and PSII changes so that there is aggregation of like molecules and complexes and consequently less capacity for transfer of excitation from LHCII to PSII and which results in dissipation of energy as heat, referred to as **non-photochemical quenching**, utilisation of light energy in production of activated electrons is **photochemical quenching**. Clustering of LHCII is increased by zeaxanthin produced by enzymatic conversion from violaxanthin.

(ii) **Reduction and repair of PSII complexes. PSII** is continually damaged, i.e. **photodamage**, and repaired, however the repair process requires movement of the PSII complex through grana to stroma thylakoids where a central protein of the complex is replaced. This process reduces the number of PSII available and so reduces flow of excited electrons, and is a component of **photoinhibition**.

(iii) **ATP production by CET.** PSI is embedded in grana thylakoids as are ATP synthase structures. Change in the balance of light harvesting complexes (see (i)) can increase activity of PSI relative to PSII and so increase ATP production required in the PSII repair system, but may restrict CO_2 fixation, and so is a component of **photoinhibition**.

(iv) **Photorespiration and the C_2-cycle.** RuBisCo can act as both a carboxylase, fixing carbon, and an oxygenase, when the 5-carbon RuDP produces one molecule each of 2-Phosphoglycolate and 3-Phosphoglycerate and a molecule of CO_2. The phosphoglycolate is processed by the **C2-cycle**, see **peroxisome**, to return glycerate to the CBB cycle in the chloroplast. Photorespiration maintains utilisation of NADPH+ and ATP produced by the light reactions of so utilises the energy of activated electrons.

(v) **Maintenance of active RuBisCo.** The active sites of RuBisCo can become blocked, for example by a sugar phosphate molecule, which is a regular process in operation of the enzyme. The process of **induction** for RuBisco involves re-activation of such sites which is achieved by thr closely associated enzyme (**rca**). The process of blocking of the active sites of RuBisCo has the advantage of reducing its susceptibility to proteolysis.

There have been many studies of photosynthesis of foliage by measuring gas exchange using cuvettes that enclose foliage and through which air is passed and CO_2 exchange between foliage and air is estimated from measurements of air flow rate and difference in CO_2 concentration before entering and leaving the cuvette. Varying illumination to the foliage while measurements are made enables calculation of an A/Q curve (A=photosynthetic rate, Q=visible radiation flux) and an A/Ci curve (Ci = CO2 concentration) by manipulating CO2 concentration. Estimates can then be made of photosynthetic potential, i.e. maximum value of A and, when change in water vapour concentration of the air flow is made, of **stomatal conductance**, gas flow through stomata, and gas passage through foliage mesophyll, i.e. **mesophyll conductance**.

photodamage: see photosynthesis.

photoinhibition: see photosynthesis.

photoreceptors: see phytochrome.

photorespiration: see photosynthesis.

photosynthetic capacity: see photosynthesis.

photosystem: see photosynthesis.

phototropism: see form.

phyllogeny C7: The evolutionary relationships among organisms illustrating lineages between organisms.

phyllotaxis: see form.

phytochrome C 8: Phytochrome is a protein, linked to a chromophore, and which acts as a photoreceptor. It is sensitive to red and far-red light that cause it to have two forms, P_r and P_{fr}. P_{fr} is produced by red light and is the physiologically active form. It is transported to the nucleus where it affects gene transcription. P_{fr} may also have affects in the cytosol. There are five types of phytochrome each affecting specific processes.

phytohormone: Molecules that are involved in controls of plant growth and development but are small and exist in low concentrations relative to molecules involved in metabolic conversions that produce the body of the plant. They may act as signal molecules to trigger a physiological response to an environmental change, or may conditioncells to respond to physiological processes. Phytohormones may act locally to where they are produced but can, in some cases, be transported over considerable distances. Major classes of phtohormones include: **abscissic acid** which generally acts as an inhibitor of growth and growth commences as its concentration decreases whilst that of **gibberellic acid** increases. **Gibberellins** promote cell elongation and transitions between vegetative and reproductive growth; **cytokinins** positively affect cell division.

phytomeres: see architecture.

plagiotropic: see form.

plasmodesmata C9: cytoplasmic connections through microscopic openings in cell walls and continuous between neighbouring cells, enabling transport of molecules between cells.

plasticity: see canopy.

pluripotent: see meristem.

proleptic: see form.

protection: see photosynthesis.

proteolysis C6: breakdown of proteins into peptides and amino acids, typically through the enzymatic action of proteases in cells.

Proximal: see form.

PSI and PSII, C3, see photosynthesis.

quantitative trait loci C1: Quantitative traits are those that vary continuously rather than being binary and are typically controlled by multiple genes and are termed **polygenic**. A **quantitative trait locus (QTL)** is a region of DNA associated with such a trait.

quenching: see photosynthesis.

radiation: see microclimate.

ramet: see clone.

redox: see photosynthesis.

regulator, C4, see Control processes

repair: of photosynthesis system, see photosynthesis

resource acquisition plasticity: see canopy.

RuBisCo: see photosynthesis.

RuBisCo activase, (rca): see photosynthesis.

scientific methods: The methods of investigation used in science influence what can be discovered and how it is expressed. Generally scientific investigations are some part of a continuous discourse between theory about a process and measurements or observations of it. Investigations with an emphasis on theory are **rationalist**, those with an emphasis on data are **empirical** but the two approaches do depend upon each other: theories are constructed about some phenomenon that is observed or inferred; empirical investigations require that a choice is made about what should be measured and this is influenced by conceptions of the functioning of a phenomenon, some theory about it.

At the early stages of investigation scientists who favour a pre-dominantly rationalist approache my investigate, or assert, the appropriateness of simple models, as in the **carbon centric approach** to the study of plant growth or in the construction of simple **growth** curves. Empirical approaches may use **correlation analysis** to investigate relationships between multiple variables, as in construction of a **multiple regression tree**.

The success of a theory may be judged by its dependence upon **causal processes** where components are assessed by their dependence on transfer of materials or energy, or their initiation and/or continuation through and information sequence. This may result in integration of different relationships to construct a **causal network**. Such networks may have parallel pathways.

seasonality: see development.

semelparous C10: reproducing in a single event prior to death.

senescence: see ageing.

set-point: control processes.

shade avoidance: see canopy.

shade tolerant: see canopy.

shoot apical meristem (SAM): see meristem.

signal type: see control processes.

sink limited photosynthesis C10: when photosynthesis is limited by the availability of growing tissue to receive its products.

source limited photosynthesis C10: when photosynthesis is limited by supply of resources, e.g. CO_2, light and water, and necessary conditions, e.g. temperature range.

spatial equalisation: see competition, C7.

stomatal conductance: see photosynthesis.

stroma: see chloroplast.

sylleptic: see form.

symplastic transport (C10): movement of molecules through the cytoplasm of cells and their interconnecting plasmodesmata.

sympodium: see form.

thylakoid: see chloroplast.

tonoplast: see cell.

upregulation: see gene expression.

quantitative model: Many types of quantitative models have been written to define plant growth or its components. Generally, models with few parameters, e.g. of the **growth curve** type (C1) have been found inadequate, although this depends upon the approach used in model assessment. Models of interactions between component processes of growth, such as **structural-functional plant models (FSPM)** have to define both component processes and their interactions. Such models typically represent some view of a **causal network**. In making the bounding decisions about such models, i.e. what should be included and how it should be represented, then assumptions have to be made. Assessment of the effects of these assumptions should be part of the overall assessment of the model.

water potential C9: the potential energy of energy per unit volume relative to that of pure water in unconstrained conditions. Pure water has a water potential of zero whereas that in plant cells has the negative value of the force required to extract the water from the plant tissue. Within the plant water tends to move from greater water potentials, i.e. less negative, to water with lesser water potentials.

whorl: see form.

Index